HYDROIDS OF THE PACIFIC COAST OF CANADA AND THE UNITED STATES

HYDROIDS OF THE PACIFIC COAST OF CANADA AND THE UNITED STATES

BY

C. McLEAN FRASER

Head of the Department of Zoology
The University of British Columbia

TORONTO
THE UNIVERSITY OF TORONTO PRESS
1937

HYDROIDS OF THE PACIFIC COAST OF CANADA AND THE UNITED STATES

BY

C. McLEAN FRASER

INTRODUCTION

In a paper, "The Hydroids of the West Coast of North America", 1911, a complete list of species known to exist on the west coast of North America, with a complete synonymy for each species, as far as Pacific coast references are concerned, was given. Included with this list was the fully recorded distribution on the coast, of each species.

Immediately after this paper appeared, extensive hydroid collecting was done in the Vancouver island and Puget sound region. As a result, "Some Hydroids of the Vancouver Island Region" was published in 1914. Additions were made to the records in this area in "Monobrachium parasitum and other West Coast Hydroids", 1918, and in "A new Hydractinia and other West Coast Hydroids", 1922.

In 1915, Volume III of Nutting's Monograph on American Hydroids, "Campanularidae and Bonneviellidae", was published, but this gave few new records for the coast.

During the years 1912 and 1913, the biological and hydrographical survey of San Francisco bay added greatly to the collected hydroid material of the coast. This *Albatross* collection, together with much other hydroid material, collected from time to time by the staff in Zoology of the University of California, was made available for examination, through the kindness of Dr. C. A. Kofoid, and under his persistent urge, the examination was completed.

The material provided such a wealth of new records from San Francisco bay, and, to a more limited extent, from other parts of the coast, that it was suggested that, since there was no general paper giving description and figures of Pacific Coast Hydroids, this would afford a good opportunity to prepare such a paper. All the new species and all the species new to the coast, in this material, were reported in "Some new and some previously unreported Hydroids, mainly from the Californian Coast", 1925.

5

Apparently the preparation of such a comprehensive paper on Pacific hydroids induced a new activity in hydroid collecting on the coast. In the summer of 1932, the staff of the Oceanographic Laboratories of the University of Washington, made an oceanographic trip from Seattle to Skagway, Alaska, on the *Catalyst*, during which some hydroids were collected, mainly in southeast Alaska. Although the amount of material appeared small, 38 species were represented. Since little collecting has been done in this area, the new distribution records help to connect up, in the hydroid realm, western Alaska with the coast farther south.

About the same time, the Biological Survey of the United States sent along hydroid material from birds' stomachs, and much of this was from the Pacific area. This was reported in "Hydroids as a food supply", 1933.

In the summer of 1934, E. G. Hart, collecting for the Biological Board of Canada from the Canadian Hydrographic Survey ship, *Wm. J. Stewart*, on the west coast of Vancouver island, included hydroids in his collecting. The paper, "Hydroids from the west coast of Vancouver island", 1935, included the records of these together with all previous records from that part of the coast.

In 1935, the opportunity offered to represent the Biological Board of Canada in collecting from the *Wm. J. Stewart* in the vicinity of the Queen Charlotte islands, mostly on the west coast of the islands, provided a means of investigating an area that, zoologically speaking, had scarcely been touched. Much material of interest was obtained, the hydroid collection being a very creditable one. Two papers on this material have appeared: "Hydroids from the Queen Charlotte islands", 1936, and "Hydroid distribution in the vicinity of the Queen Charlotte islands", 1936.

Mr. E. F. Ricketts of Pacific Grove, in his visits to various parts of the coast to obtain biological supplies, has picked up hydroids on several occasions, and these have provided some distributional records of interest.

Last, but not least, a collecting trip on the *Catalyst* with some of the staff of the Oceanographic Laboratories of the University of Washington, in the summer of 1936, although of only ten days' duration, was very profitable in this regard. Collecting was done at the north end of Vancouver island, in Queen Charlotte sound; in Hecate strait; in the Queen Charlotte island area—Houston Stewart channel, west coast of Moresby and Graham islands, Dixon entrance, across the north end of Graham island to Rose spit, along the east side of Graham island to Skidegate inlet, and on through Hecate strait in a direct course from Skidegate to the entrance to Queen Charlotte sound. Besides the addition of numerous distribution records to known species, the hydroid material collected provided a new genus and new species, two species not previously reported from the Pacific coast and two new gonangia. These are reported and described in this paper.

At this stage, the paper has been completed. It is an attempt to give a brief description, with figures, of every hydroid species known to occur along the Pacific coast of Canada and the United States, together with its distribution within this area. It is intended to provide the Pacific zoologist with a reference, easily understood, to every species of hydroid reported from the coast. Keys to families, genera, and species have been included to facilitate diagnosis. Much of the information presented has already been published, but in widely scattered papers, some of them long out of print.

The new contribution is largely in the extensive addition to the distribution records, for which many thousands of specimens have been examined. These new records have been provided by the *Albatross* material and that which accompanied it from the University of California at Berkeley; the material from the *Catalyst* Seattle-Skagway expedition; the material collected by E. F. Ricketts in many locations along the coast; the material, mostly dredged, from the *Catalyst* Vancouver island-Queen Charlotte islands collecting trip.

A new genus and a new species, *Tetranema furcata*; four new gonangia, those of *Clytia hesperia, Halecium densum, Selaginopsis cylindrica*, and *Sertularella tanneri* and two species new to the coast, *Selaginopsis alternitheca* and *Thuiaria lonchitis*, are recorded and described.

As to the paper in general, where material was available, descriptions and figures were made from the actual specimens, either directly for this paper, or earlier for some previous paper. For some species that have been reported from the coast, no material was available for examination, hence, for these species, descriptions and figures published by other authors were used. In all such cases direct reference is made to the source. For three species included, *Hydractinia californica, Corymorpha carnea*, and *Tubularia borealis* no figures could be obtained. Apparently none was ever published

Unless there was some special reason, such as great size, for an exception (in which case, the magnification is indicated in the legend), all the enlarged drawings are made to the same scale of magnification (20 diameters), hence it was not considered necessary to give specific measurements.   Since the general appearance of a hydroid colony often helps greatly in diagnosis of the species, natural size drawings have been provided in all possible cases.

Without the figures the paper would lose much, if not most, of its value for reference.   My wife has made the drawings and it pleases me greatly to be able to acknowledge her contribution.

Acknowledgements have been made in the various papers that have been published, of the assistance and co-operation by numerous individuals, but this is the first opportunity that has arisen to express publicly my obligation to the United States Bureau of Fisheries, for giving permission to make use of the *Albatross* collections; to the Department of Zoology, University of California at Berkeley, through Dr. C. A. Kofoid, for placing the extensive collection in the Department in my hands for examination; to the staff of the Oceanographic Laboratories of the University of Washington, for providing material and means of obtaining it; to Mr. E. F. Ricketts of Pacific Grove for his contributions at various times, and to any others who have contributed material or given assistance in other ways to make this paper as it is, possible.

Even with all of this assistance and co-operation, the work might have remained in oblivion if the National Research Council of Canada had not come to the rescue by agreeing to sponsor the publication, hence the final expression of obligation rests with the Research Council.

Geographic Distribution

In considering geographic distribution of the hydroids of the Pacific coast, a distribution table is presented to show the distribution, throughout the whole of this region, of each species that has been reported, or is now reported, from the Pacific coast of Canada and the United States.   No attempt is made to relate this distribution with the general distribution of each species.   That is a large proposition in itself, that may be considered at some later date.

For the purpose of preparing the table, the whole coast has been divided, for convenience, into six areas, although these are not equal in extent, nor are they delimited through any lack of continuity in distribution.

The first area includes that portion of the coast that lies between the United States-Mexican boundary and San Francisco bay. In this area, inshore collecting has been rather extensive and somewhat isolated collections have been obtained from deep water. How representative these collections are it is impossible to say.

The second area includes all of San Francisco bay and the shallow water area outside the Golden Gate, extending somewhat to the westward of a line running from point Lobos to Bonita point. This is entirely a shallow water area; it has been extensively worked by the University of California and, more particularly, by the scientific staff of the United States Bureau of Fisheries steamship *Albatross*, in the survey of this region in 1912-1913. It is not a large area as compared with the whole coast, but it makes quite a definite unit in the contribution to the knowledge of the Pacific coast hydroids.

The third area extends from San Francisco bay to cape Flattery. Little shore collecting of hydroids has been done in this area, and although the *Albatross* did some dredging off the coast in deeper water, collections have been examined from very few stations. It is a large and promising area, but, at the present time, less is known of the hydroid possibilities here than in any other area on the whole coast.

The fourth area includes the Vancouver island and Puget sound regions. In a portion of the area—in the Nanaimo region and in the San Juan archipelago—intensive collecting has been done. Previous to 1934, the west coast of Vancouver island had received little attention, but in that year, the collecting done by E. G. Hart for the Biological Board of Canada, added materially to the distribution records. Now, most of the area has received some attention, although much of it would be the better of further exploration.

The fifth area includes all of British Columbia outside of area number IV. Apart from the region of the Queen Charlotte islands, this area has scarcely been touched. The collections from the vicinity of the Queen Charlottes in 1935 and 1936 have quite creditably rescued this portion from oblivion.

The sixth area includes the whole Alaskan coast, and it is therefore larger than any of the other areas. Expeditions such as the Harriman expedition, the *Albatross* expedition, expeditions of the Coast and Geodetic Survey to the gulf of Alaska, the Aleutian islands, and the Bering sea, and the expedition of the *Catalyst* to southeast Alaska, have obtained extensive collections from this area, but, in the main, these have been of dredged material. The inshore and shallow water material has not been well represented.

In areas II, IV, and V, shore and shallow water, or comparatively shallow water, species predominate; in areas III and VI, deeper water forms are in the majority, and in area I they are more nearly balanced.

In spite of such obvious deficiencies in available material, the records show a decided trend towards continuity in distribution, rather than that there is any indication that there is any general, abrupt cessation at any point along the coast; 236 species are included in the distribution table. Of 86 species in area I, only 15, 17%, are restricted to the area; of 93 in area II, only 13, 14%; of 41 in area III, only 1, 2%; of 144 in area IV, only 18, 12%; of 105 in area V, only 12, 11%; of 119 in area VI, only 28, 24%. Cape Flattery may seem to be a distinct dividing point, but yet, of 128 species from south of this point, only 42, 33%, are restricted to the south, and of 192, only 109, 57%, are restricted to the north.

In general, the continuity of distribution is shown in each of the families represented on the coast. The accompanying table shows the numbers in each family, or group of families, as well as the percentage that each number bears to the total number of species in that family or group, from the whole coast.

DISTRIBUTION TABLE

	Total	I No.	I %	II No.	II %	III No.	III %	IV No.	IV %	V No.	V %	VI No.	VI %
Gymnoblastea ...	49	17	35	19	39	7	14	28	57	14	29	18	37
Bonneviellidae.. } Campanularidae }	49	22	45	24	49	9	18	32	65	27	55	22	45
Campanulinidae..	9	1	11	4	44	1	11	9	100	3	33	3	33
Halecidae.......	25	6	24	9	36	2	8	15	60	11	44	13	52
Hebellidae..... } Lafoeidae..... } Synthecidae... }	13	6	46	7	54	1	8	9	69	7	54	10	77
Sertularidae.....	69	18	26	23	33	14	20	40	58	37	54	46	67
Plumularidae....	22	16	73	7	32	7	32	9	41	8	36	6	27

In each area the percentages are quite uniform throughout, with two exceptions. The family Plumularidae is especially well represented in area I, while it shows marked deficiency in areas IV, V, and VI, and the Lafoeidae and Sertularidae are especially well represented in area VI. This follows the general rule that the Plumularidae is much richer in species in tropical and sub-tropical areas than in higher latitudes, while the Sertularidae, particularly in the genera represented to the greatest extent on this coast, is more extensively varied in the colder waters.

The Gymnoblastea is not very well represented in the records in any of the areas except IV. There may be a definite explanation for this. On account of a partial or a complete absence of perisarc, the gymnoblastic individuals or colonies are very easily injured or destroyed. In dredged material, many of them may be almost unrecognizable even when they reach the deck. If they are preserved *en masse* with other material, the last chance of recognition often disappears. A sorting out of the hydroids by themselves and preservation as soon as possible, often provides the only chance of saving material for examination. In much of the collecting, provision cannot be made for any such special attention.

DISTRIBUTION TABLE

	I	II	III	IV	V	VI
Clava leptostyla............		x				
Corydendrium fruticosum...				x		
Endocrypta huntsmani......				x		
Tubiclava cornucopiae......					x	
Turris neglecta............					x	
Turritopsis nutricula........		x				
Monobrachium parasitum...				x	x	
Coryne brachiata..........						x
Coryne corrugata..........	x					
Coryne crassa.............				x		
Syncoryne eximia..........	x	x				x
Syncoryne mirabilis........	x	x	x	x	x	x
Atractyloides formosa.......	x					
Bimeria franciscana........		x				
Bimeria gracilis...........	x	x	x	x	x	
Bimeria pusilla............		x				
Bimeria robusta...........	x			x		
Bimeria tenella...........		x				
Garveia annulata..........	x	x	x	x	x	x
Garveia formosa..........	x	x				
Garveia groenlandica.......				x	x	x
Garveia nutans...........						x
Bougainvillia glorietta......	x			x		
Bougainvillia mertensi......		x		x		x

DISTRIBUTION TABLE—*Continued*

	I	II	III	IV	V	VI
Perigonimus repens.........		x		x	x	x
Perigonimus serpens........					x	
Eudendrium californicum....	x	x	x	x		
Eudendrium capillare.......		x		x		x
Eudendrium insigne........				x		
Eudendrium irregulare......				x		
Eudendrium rameum.......	x	x		x	x	x
Eudendrium ramosum......	x	x				
Eudendrium tenellum.......	x			x	x	x
Eudendrium vaginatum.....			x	x	x	x
Hydractinia aggregata......				x	x	
Hydractinia californica......	x					
Hydractinia laevispina......				x		
Hydractinia milleri.........		x		x		
Pennaria tiarella..........		x				
Corymorpha carnea........						x
Corymorpha palma.........	x					
Tubularia aurea...........					x	
Tubularia borealis.........						x
Tubularia crocea..........	x	x		x		x
Tubularia harrimani.......				x		x
Tubularia indivisa.........				x		x
Tubularia larynx..........				x		
Tubularia marina..........	x	x	x	x		
Hybocodon prolifer........				x		x
Bonneviella regia..........				x	x	x
Bonneviella superba........						x
Campanularia castellata.....		x				
Campanularia denticulata...	x			x		x
Campanularia exigua.......				x		
Campanularia fusiformis.....	x		x	x		
Campanularia gelatinosa....		x		x	x	
Campanularia gigantea.....					x	
Campanularia groenlandica..	x			x	x	x
Campanularia hincksi.......	x	x				
Campanularia integra.......	x			x	x	x
Campanularia raridentata...		x		x	x	
Campanularia ritteri........	x					x
Campanularia speciosa......				x	x	x
Campanularia urceolata.....	x	x	x	x	x	x
Campanularia verticillata ...	x	x	x	x	x	x
Campanularia volubilis......	x			x	x	x
Clytia attenuata..........		x		x		
Clytia bakeri..............	x	x				
Clytia cylindrica..........				x	x	

DISTRIBUTION TABLE—*Continued*

	I	II	III	IV	V	VI
Clytia edwardsi	x	x		x	x	
Clytia hendersoni	x	x				
Clytia hesperia	x					
Clytia inconspicua				x	x	
Clytia johnstoni		x		x	x	x
Clytia kincaidi				x	x	x
Clytia longitheca		x		x		
Clytia minuta					x	
Clytia universitatis	x	x				
Eucopella caliculata		x		x	x	x
Eucopella compressa	x					x
Eucopella everta	x	x	x	x	x	
Gonothyraea clarki	x	x	x	x	x	x
Gonothyraea gracilis				x	x	x
Gonothyraea inornata				x	x	x
Obelia bicuspidata		x				
Obelia borealis				x	x	x
Obelia commissuralis		x				
Obelia corona	x	x				
Obelia dichotoma	x	x	x	x	x	x
Obelia dubia	x	x	x	x	x	x
Obelia fragilis				x		
Obelia geniculata	x	x	x		x	
Obelia gracilis				x		
Obelia griffini				x		
Obelia longissima	x	x	x	x	x	x
Obelia multidentata				x		
Obelia plicata		x		x	x	x
Obelia surcularis				x		
Calycella syringa	x	x	x	x	x	x
Campanulina forskalea		x		x		
Campanulina rugosa				x		x
Cuspidella grandis				x	x	
Cuspidella humilis		x		x		
Egmundella gracilis		x		x	x	
Lovenella producta				x		
Opercularella lacerata				x		
Stegopoma plicatile				x		x
Campalecium medusiferum	x					
Halecium annulatum	x	x	x	x	x	x
Halecium articulosum				x	x	
Halecium beani		x				
Halecium corrugatum	x	x		x	x	
Halecium densum				x	x	
Halecium flexile				x	x	

DISTRIBUTION TABLE—*Continued*

	I	II	III	IV	V	VI
Halecium halecinum........		x		x		x
Halecium kofoidi...........	x	x		x		
Halecium labrosum.........				x	x	x
Halecium muricatum.......						x
Halecium ornatum.........						x
Halecium parvulum........				x		
Halecium pygmaeum.......		x		x	x	
Halecium reversum.........				x	x	x
Helecium robustum.........						x
Halecium scutum...........				x		x
Halecium speciosum........		x				x
Halecium telescopicum......						x
Halecium tenellum.........	x	x	x	x	x	x
Halecium washingtoni......	x	x		x	x	x
Halecium wilsoni...........				x	x	x
Ophiodissa carchesium......						x
Ophiodissa corrugata.......					x	
Ophiodissa gracilis..........					x	
Hebella pocillum...........	x	x		x		x
Acryptolaria pulchella......		x				
Filellum serpens............	x	x	x	x	x	x
Grammaria abietina........				x	x	x
Grammaria immersa........				x	x	x
Lafoea adhaerens...........						x
Lafoea adnata.............	x					
Lafoea dumosa.............	x	x		x	x	x
Lafoea fruticosa............		x		x	x	x
Lafoea gracillima...........	x	x		x	x	x
Lictorella carolina.........				x		x
Lictorella cervicornis.......				x	x	x
Synthecium cylindricum.....	x	x				
Abietinaria abietina........	x	x	x	x	x	x
Abietinaria alexanderi.......						x
Abietinaria amphora........		x	x	x	x	x
Abietinaria anguina.........	x	x	x	x	x	
Abietinaria annulata........						x
Abietinaria costata.........					x	x
Abietinaria filicula..........		x	x	x	x	x
Abietinaria gigantea........		x			x	x
Abietinaria gracilis.........				x		x
Abietinaria greenei.........	x	x	x	x	x	
Abietinaria inconstans......						x
Abietinaria pacifica.........	x					
Abietinaria rigida..........				x	x	x
Abietinaria traski..........	x	x	x	x	x	x

DISTRIBUTION TABLE—*Continued*

	I	II	III	IV	V	VI
Abietinaria turgida	x	x	x	x	x	x
Abietinaria variabilis			x	x	x	x
Dictyocladium flabellum				x		x
Diphasia corniculata		x				
Diphasia kincaidi						x
Diphasia pulchra				x		
Hydrallmania distans		x		x	x	
Hydrallmania franciscana		x				
Selaginopsis alternitheca					x	
Selaginopsis cedrina						x
Selaginopsis cylindrica				x	x	x
Selaginopsis hartlaubi				x		x
Selaginopsis mirabilis				x	x	x
Selaginopsis obsoleta						x
Selaginopsis ornata					x	x
Selaginopsis pinaster						x
Selaginopsis pinnata				x	x	x
Selaginopsis trilateralis					x	
Selaginopsis triserialis	x	x		x		
Sertularella albida				x	x	x
Sertularella clarki						x
Sertularella complexa						x
Sertularella conica	x		x	x	x	x
Sertularella elegans						x
Sertularella fusiformis		x	x			
Sertularella levinseni						x
Sertularella magna						x
Sertularella minuta						x
Sertularella pedrensis	x	x		x	x	
Sertularella pinnata		x		x	x	x
Sertularella polyzonias				x	x	x
Sertularella rugosa		x		x	x	x
Sertularella tanneri				x	x	
Sertularella tenella	x			x	x	
Sertularella tricuspidata	x	x	x	x	x	x
Sertularella turgida	x	x	x	x	x	
Sertularia cornicina	x	x				
Sertularia desmoides	x	x				
Sertularia furcata	x	x		x	x	
Sertularia pumila	x					
Thuiaria alba				x	x	
Thuiaria argentea	x	x	x	x	x	x
Thuiaria carica				x		
Thuiaria dalli	x			x	x	x
Thuiaria distans				x	x	

DISTRIBUTION TABLE—*Continued*

	I	II	III	IV	V	VI
Thuiaria elegans						x
Thuiaria fabricii				x	x	x
Thuiaria kurilae						x
Thuiaria lonchitis					x	
Thuiaria plumosa						x
Thuiaria robusta				x		x
Thuiaria similis		x	x	x	x	x
Thuiaria tenera				x		x
Thuiaria thuiarioides				x	x	x
Thuiaria thuja				x		x
Aglaophenia diegensis	x	x				
Aglaophenia inconspicua	x	x	x			
Aglaophenia latirostris	x	x	x	x	x	
Aglaophenia lophocarpa	x					
Aglaophenia octocarpa	x					
Aglaophenia pluma	x			x		
Aglaophenia struthionides	x	x	x	x	x	x
Antennella avalonia	x					
Antennularia verticillata			x			
Cladocarpus vancouverensis				x	x	
Diplocheilus allmani	x					
Nuditheca dalli						x
Plumularia alicia	x	x	x			
Plumularia corrugata				x	x	
Plumularia goodei	x			x		
Plumularia halecioides					x	x
Plumularia lagenifera	x	x	x	x	x	x
Plumularia megalocephala	x					
Plumularia plumularoides	x			x		x
Plumularia setacea	x	x	x	x	x	x
Plumularia virginiae	x					
Tetranema furcata					x	

BATHYMETRICAL DISTRIBUTION

In very few instances in the earlier Pacific hydroid papers has the depth been recorded. In the new material examined for this paper, and for some of the more recent papers, the depth was included in most of the records. As far as definite information is available, the limits of distribution for each species on this coast are given, but in many cases one must surmise that these limits would be extended, more or less, if there were complete records of depth. When most of the species obtained by dredging in 60 fathoms or more, eight or nine miles off the coast of Oregon, near Heceta head, Yaquina point, *etc.*, were also obtained in various locations in San Francisco bay, particularly, in a small area near Alcatraz island, and another small area in the southern section of the bay, near Shag rock and San Bruno light, in water sometimes as shallow as two or three fathoms, it is more than probable that many other species have a wider bathymetrical range than is given here.

It is unfortunate that there is not more information because, in all probability, a study of bathymetrical distribution would be quite as interesting as one of geographical distribution.

SYSTEMATIC DISCUSSION

With few exceptions, the nomenclature used in this paper is similar to that used in previous papers on North American hydroids. The few changes made are not likely to cause any confusion.

Nearly all of the authors of recent papers on hydroids have considered it wiser to take a conservative attitude rather than introduce changes that add to the synonymy without making the classification any clearer.

In the grouping of genera into families in the Gymnoblastea, Allman's system was followed in earlier papers. No matter how similar the trophosomes in two genera, Allman placed them in different families, if, in the one sporosaes are produced, and in the other free medusae are developed. Hincks does not do so unless there is a significant difference in the trophosome as well. As no systematist hesitates to put *Clytia* and *Obelia* with *Campanularia* in the *Campanularidae*, it appears to be more nearly consistent to follow Hincks's system rather than Allman's, in the Gymnoblastea. Consequently, in this paper, the genera *Clava, Corydendrium, Endocrypta, Tubiclava, Turris*, and *Turritopsis*, are all placed in the family *Clavidae*; *Coryne* and *Syncoryne* in the family *Corynidae* and *Atractyloides, Bimeria, Bougainvillia, Garveia*, and *Perigonimus* in the family *Atractylidae*.

In the Calyptoblastea, Nutting has been followed in separating the genus *Bonneviella* from the *Campanularidae* to establish the family *Bonneviellidae*, since it is considered that the presence of a preoral cavity, bounded internally by a veloid, is a character significant enough to justify the separation. In the family *Campanulinidae*, the genus *Egmundella*, established by Stechow, has been included. The genus *Hebella* has been placed, as formerly, in the

2

family *Hebellidae*. In the *Lafoeidae*, Nutting and others have called attention to the fact that the genus, formerly called *Cryptolaria*, should be *Acryptolaria*, and that change has been made. The genus *Synthecium* does not fit in with the *Sertularidae*, since there is no opercuium on the hydrotheca and the pedicellate gonangia arises from the interior of the hydrotheca. It must be placed in a family apart, the *Synthecidae*, as Stechow has indicated.

It is in the *Sertularidae* that Stechow attempts changes in nomenclature to the greatest extreme. In his 1923 paper, "Zur Kenntnis der Hydroidenfauna des Mittelmeeres, Amerikas und anderer Gebiete", he discards most of the old established generic names, and introduces so many new ones, that it would seem that it is his ultimate aim to have a genus for each species. Happily, no one has followed his lead to any great extent, and in consequence, he has a synonymy all his own. Kramp is surely justified when he states in his 1932 paper, "The Godthaab Expedition Hydroids", that there is no real basis for discarding "*Thuiaria*" for "*Salacia*" or "*Diphasia*" for "*Nigellastrum*", and he might well have added others to the list.

Intergrading forms have often given rise to heated discussions, and this applies to hydroid systematists as well as those in other groups. Such discussions must be of little avail. When new species, genera, or families, are being continually differentiated, we must expect intergrading forms. Why not accept them as such instead of trying to force them in with their closest relatives?

As a definite case in point, the species here called *Thuiaria thuiarioides* may be cited. This is not a typical *Thuiaria* because the hydrotheca has an operculum of one adcauline flap, instead of one abcauline flap. If the systematic position were decided on that character alone, it should be an *Abietinaria* or a *Diphasia*, but the shape of the hydrotheca, the shape of the gonangium, its habit of growth, and other features are distinctly thuiarian. For that reason it is placed with *Thuiaria* rather than with either of the other genera. As long as it is recognized that it is an intergrading form, it makes little difference, though, whether it is called *Thuiaria*, *Abietinaria*, or *Diphasia*.

In the 1911 paper, "The Hydroids of the West Coast of North America", there was included with each species listed, complete synonymy as far as the papers dealing with the hydroids of this coast were concerned. There is nothing to be gained by repeating all of that synonymy here. In each case, where the species was listed in that paper, a page reference is given. If the species has been included in any later west coast paper, a reference to this is also given. The only extensive paper dealing exclusively with west coast hydroids since 1911, is the 1914 paper, "Some Hydroids of the Vancouver Island region". In this, all the species listed are described and figured, hence frequent reference is made to it. The reference to the original description is given for each species and in the case of species that are not mentioned in either of the above-mentioned papers, one or more references are given to add effectiveness to the account of the species.

Sub-order GYMNOBLASTEA

Hydroids with hydranths unprotected by hydrothecae and gonophores unprotected by gonangia or other structures serving a similar purpose.

KEY TO THE FAMILIES

A. Hydranths with filiform tentacles only
    a. Hydranths with scattered filiform tentacles......*Clavidae*
    b. Hydranths with a single basal whorl of filiform tentacles
        a. Colonies without perisarc on the zooids..*Hydractinidae*
        b. Colonies with perisarc
            I. Hydranth with conical proboscis.....*Atractylidae*
            II. Hydranth with trumpet-shaped proboscis ....
            ....*Eudendridae*
    c. Hydranth with but one tentacle.........*Monobrachiidae*
    d. Hydranth with proximal and distal tentacles
        a. Distal tentacles in one whorl...........*Tubularidae*
        b. Distal tentacles in two whorls........*Hybocodonidae*
        c. Distal tentacles in several whorls......*Corymorphidae*
B. Hydranths with scattered capitate tentacles; no filiform tentacles.....................................*Corynidae*
C. Hydranths with a basal whorl of filiform tentacles and distal scattered capitate tentacles................*Pennaridae*

Family Clavidae

Trophosome.—Hydranths with scattered filiform tentacles.

Gonosome.—Gonophores giving rise to sporosacs or to free medusae.

KEY TO THE GENERA

A. Gonophores produce sporosacs
  a. Zooids arising singly from a stolon
    a. No rigid perisarc............................*Clava*
    b. Perisarc rigid............................*Tubiclava*
  b. Colony fascicled and branched............*Corydendrium*
B. Gonophores produce free medusae
  a. Zooids arising singly
    a. Stoloniferous network degenerate and covered with an
        encrusting perisarc.................*Endocrypta*
    b. Zooids arising from a filiform stolon, perisarc poorly
        developed, if at all.....................*Turris*
  b. Colony branched.........................*Turritopsis*

Genus CLAVA

Trophosome.—Zooids arising singly from a reticular stolon; tentacles numerous, scattered, filiform; proboscis clavate.

Gonosome.—Gonophores produce fixed sporosacs in clusters a short distance below the proximal tentacles.

Clava leptostyla Agassiz

Plate 1, Fig. 1

Clava multicornis STIMPSON, Marine Invert. Grand Manan, 1853,
          p. 16.
Clava leptostyla AGASSIZ, Cont. Nat. Hist. U.S., Vol. IV, 1862,
          p. 218.
         HINCKS, Br. Hyd. Zooph., 1868, p. 6.
         FRASER, Can. Atlantic Fauna, Hydroida, 1921, p. 7.

Trophosome.—Zooids clustered, 1 cm. in height, constricted at the base; clavate; tentacles 20-30.

Gonosome.—Sporosacs spherical, appearing in large clusters just below the proximal tentacles.

Distribution.—In much of the shore line, shallow water—10 fathoms or less—of San Francisco bay; from Mare island to Oakland and in the lower section of the bay as far as San Mateo. Off Lime point is the farthest seaward location noted.

The distribution of this species is interesting. It is a common species off the coast of New England, New Brunswick, and Nova Scotia. There is nothing to indicate continuous distribution to San

Francisco bay, either by a northern or a southern route. Possibly it has been transported by boat, and, from a focus in the bay, has spread into suitable areas in all parts of the bay.

Genus CORYDENDRIUM

Trophosome.—Stems fascicled and branched; hydranths with scattered filiform tentacles.

Gonosome.—Gonophores borne on stems or branches, producing fixed sporosacs.

? Corydendrium fruticosum Fraser

Plate 1, Fig. 2

Corydendrium fruticosum FRASER, Hyd. Vancouver island region, 1914, p. 112.

Trophosome.—Colony consisting of a short, thick, much fascicled stem, which gives rise to a tuft of rather long branches of nearly the same length, up to 3 cm. These branches seldom give off other branches but pass out almost straight, fascicled more or less throughout and decreasing very little in size. The hydranths, which are supported on short pedicels, usually come off in nearly opposite pairs but sometimes they appear singly. These pairs are seldom far apart but become more crowded distally. The thick perisarc forms a heavy tube around the pedicel of the hydranth so that the end of it is very noticeable. The hydranth when fully developed is large and is provided with 12 to 15 tentacles that have no very definite arrangement.

Gonosome.—Unknown.

Distribution.—East of Brown island, Friday Harbor, 40-60 fathoms, Snake island, 20-30 fathoms, Northumberland channel, 20-30 fathoms (Fraser).

Genus ENDOCRYPTA

Trophosome.—Hydrorhiza of small fibres or almost entirely degenerated. The perisarc that envelops the hydrocaulus also unites one with the others, this being in the nature of a very fine encrustation. Hydranths clavate.

Gonosome.—Gonophores producing free medusae with four radial canals and four simple marginal tentacles.

Endocrypta huntsmani (Fraser)

Plate 1, Fig. 3

Crypta huntsmani FRASER, West Coast Hydroids, 1911, p. 19.

Endocrypta huntsmani FRASER, Science, Vol. XXXV, 1912, p. 216.
                    Hyd. from Vancouver I., 1913,
                    p. 149.
                    Hyd. of the V.I. region, 1914, p. 109.

Trophosome.—Hydrocaulus tubular, in most cases erect and unbranched, but occasionally it branches or forks, as the two parts seem to be of equal significance and seldom differ much in size. The two parts of the fork may remain in an upright position forming a very acute angle, or one or both of them may turn much from the perpendicular and the one may turn from the other until they get in exactly opposite directions. The erect hydrocaulus may reach a height of 8 mm. The perisarc is so thin as to be a mere pellicle, that part which forms the basal expansion being particularly so. The hydranth appears much darker than the hydrocaulus; when at rest it is club-shaped but it has extreme mobility. The mouth may be opened so that the proboscis is trumpet-shaped or it may be averted and reflexed entirely so that it is folded back over the base of the tentacles, these also being turned back to point towards the base. Tentacles up to 24 in number, with seldom any definite arrangement.

Gonosome.—Medusa buds, from 1 to 3, usually 2, are developed a short distance below the tentacles. The medusae are about 2 mm. in diameter at time of liberation.

Distribution.—Departure bay; Nanoose bay; near Clarke rock; off Protection island, Friday Harbor (Fraser). In all cases they were obtained from the branchial cavity of certain species of ascidians, attached to the peribranchial wall.

Genus TUBICLAVA

Trophosome.—Colonies erect, simple or branched; the perisarc forms a stout tube around the whole hydrocaulus, into the end of which the hydranth may be retracted. Hydranths clavate, with scattered filiform tentacles.

Gonosome.—Gonophores producing fixed sporosacs in clusters grow either from the stolon or the hydrocaulus.

Tubiclava cornucopiae Norman
Plate 1, Fig. 4

Tubiclava cornucopiae NORMAN, Ann. and Mag. N.H. (3) XIII, 1864, p. 82.

HINCKS, Br. Hyd. Zooph., 1868, p. 11.

FRASER, Hyd. from the Queen Charlotte Is., 1935, p. 505.

Trophosome.—Nutritive zooids growing singly but rather closely segregated, from a stolon that forms a loose network over shells of living bivalve molluscs; up to 5 mm. in length. The perisarc forms a trumpet-shaped tube which is slightly curved and may be reduplicated; hydranths clavate, quite extensile, with 16 or more scattered tentacles.

Gonosome.—Sporosacs in mulberry-like masses are supported on short pedicels, surrounded by perisarc in much the same way as the hydrocaulus; scattered about among the nutritive zooids.

Distribution.—Outside Massett inlet, 15 fathoms (Fraser).

Genus **TURRIS**

Trophosome.—Zooids growing singly from a stolon, pedicels short; hydranth clavate with scattered filiform tentacles.

Gonosome.—Gonophores are unknown but free medusae are produced.

? Turris neglecta Lesson
Plate 1, Fig. 5

Turris neglecta LESSON, Hist. Nat. Zooph., 1843, p. 284.

ALLMAN, Ann. and Mag. N.H., (3), XIII, 1864, p. 352.

HINCKS, Br. Hyd. Zooph., 1868, p. 13.

FRASER, Hyd. from Q.C.I., 1935, p. 505.

Trophosome.—Zooids growing singly from a filiform stolon; perisarc thin, forming a pellicle; hydranths clavate with 10 or more scattered tentacles.

Gonosome.—Unknown. Hincks describes the medusa as follows: "Gonozooid free and medusiform. Umbrella sub-cylindrical, with 4 or 8 longitudinal bands; manubrium massive, four-lipped; radiating canals 4; marginal tentacles numerous, each with an ocellated bulbous base".

Distribution.—North shore of Hibben island, on *Pugettia gracilis*, low tide (Fraser).

Genus TURRITOPSIS

Trophosome.—Small colonies with few branches from a much branched stolon.   Perisarc reaching to the base of the hydranth.

Gonosome.—Gonophores give rise to medusae with four radial canals and several simple marginal tentacles.

Turritopsis nutricula McCrady
Plate 1, Fig. 6

Turritopsis nutricula McCrady, Gymnoph.   Charleston har., 1857, p. 25.

Fraser, Hydroids of Beaufort, N.C., 1912, p. 345.

Trophosome.—Mature colony slightly branched, each branch bearing a single hydranth.   Perisarc thick, ending abruptly immediately below the hydranth.   Proboscis clavate, elongated.   Tentacles arranged in a series of somewhat irregular rows.

Gonosome.—Gonophores, each giving rise to a single medusa, appear on short pedicels at the base of the hydranth.   Each medusa-bud is invested with perisarc.   At the time of liberation the medusa has eight tentacles, but the number is greatly increased later.   The mature medusa has a quadrate stomach and a four-lipped mouth.

Distribution.—In San Francisco bay, probably near Oakland. It is a shallow water or littoral species.

Family Monobrachiidae

Trophosome.—Zooids growing individually from a stolon; hydranth with only one tentacle.

Gonosome.—Gonophores producing free medusae.

Genus MONOBRACHIUM

Trophosome.—Zooids without perisarc, growing singly from a reticular stolon; hydranth with but one tentacle.

Gonosome.—The gonophore arising from the stolon produces a single medusa with four radial canals.

Monobrachium parasitum Mereschkowsky
Plate 2, Fig. 7

Monobrachium parasitum Mereschkowsky, Hydroids from the White Sea, 1877, p. 226.

LEVINSEN, Meduser, Ctenophorer og Hydroider fra Groenlands Vestkyst,
1893, p. 151.
FRASER, Monobrachium parasitum and
other west coast hyd., 1918, p. 131.

Trophosome.—Colony consisting of many zooids, about 20,
growing from a reticular stolon which appears on the surface of
small living shells. The zooids appear at the hinge of the shell but
the network extends over the surface to the margin, with a number
of free ends projecting beyond it. The terminal portion of these
free ends consists of a globular mass of large thread cells, or cells
having a similar appearance, having no perisarcal protection. They
may be defensive zooids. The individual zooid is small, 0.7 to
0.8 mm. in height, fusiform, possessing great freedom of movement;
the proboscis is approximately half as long as the remainder of the
zooid; the mouth is terminally placed; around it thread cells are
closely arranged but they are not in groups; further down on the
proboscis they are much less numerous; there is no constriction at
the base of the proboscis. The one tentacle comes out almost at
right angles to the body but except for the basal portion no definite
position can be given for it as it contracts and expands regularly and
moves about freely; when extended, it is longer than the body,
about 1 mm.; the surface is well provided with thread cells but these
are scattered singly over the whole length.

Gonosome.—The medusa-buds grow from the stolon, supported
by a very short pedicel; they are invested with a pellicle. The number of buds in any one colony seems to be rather limited, never more
than four in any of the colonies obtained, two being the usual number. Only one medusa is produced from each gonophore; it is almost globular, with four radial canals.

Distribution.—Nanoose bay, 10-15 fathoms, on living shells of
the bivalve, *Axinopsis sericatus* (Carpenter) (Fraser); Rennell
sound, 60-70 fathoms (Fraser).

Family Corynidae

Trophosome.—Perisarc well developed; hydranths clavate with
scattered capitate tentacles.

Gonosome.—Gonophores producing sporosacs or free medusae.

Genus **CORYNE**

Trophosome.—Colony branched or unbranched; hydrorhiza of creeping filiform tubes; the stems and branches invested with a thick perisarc. Hydranths clavate with scattered capitate tentacles.

Gonosome.—Sporosacs developed from the body of the hydranth, among, or just proximal to, the tentacles.

Coryne brachiata Nutting
Plate 2, Fig. 8

Coryne brachiata NUTTING, Hyd. of the Harriman Expedition, 1901, p. 165.

TORREY, Hyd. of the Pacific coast, 1902, p. 8.

FRASER, West Coast Hyd., 1911, p. 21.

Trophosome.—"Colony forming a dense tuft of irregularly branching stems, sometimes attaining a height of about three-quarters of an inch. Stems and branches profusely and regularly annulated throughout, fairly stout except at the proximal ends where they taper gradually to their point of origin; distal ends of many of the branches bear a more or less regular whorl, or radiating cluster, of annulated branchlets, just below the hydranth body, reminding one of the whorls of cirri around the stem of the stalked crinoids. Hydranths large, with long, slender body and proboscis, and numerous (20-35) capitate tentacles arranged in a scattered or sub-verticillate manner, over nearly the whole surface" (Nutting).

Gonosome.—"Gonophores very numerous, borne among the tentacles on the hydranth bodies, globular in outline and showing no trace of radial canals or other medusoid structures" (Nutting).

As no specimens are available for description, Nutting's description and figures are given.

Distribution.—Yakutat bay, Alaska (Nutting).

? Coryne corrugata Fraser
Plate 2, Fig. 9

Coryne corrugata FRASER, Some new and some previously unreported Hyd., 1925, p. 167.

Trophosome.—Colony reaching a height of 3.5 cm., much branched, the branches coming off with a definite knee-joint at a very acute angle with the stem; the secondary branches arise from the primary branches in the same manner; many of these secondary branches, while terminating in a hydranth, give off numerous branchlets that have no hydranths; perisarc thick, with deep annulations, that may be considered as corrugations, throughout the whole length of the stem and branches; hydranths capable of great extension and usually appearing long and slender; tentacles 20-30, not arranged in very perfect verticils.

Gonosome.—Unknown.

Distribution.—San Diego, near jetty (Fraser).

Coryne crassa Fraser
Plate 2, Fig. 10

Coryne crassa FRASER, Hyd. of the V.I. region, 1914, p. 113.

Trophosome.—Colony slightly and irregularly branched, branches reaching a height of 15 mm. The branches come from the stem at a wide angle and are constricted at the base; stem and branches practically the same size throughout; perisarc thick with but few annulations. Hydranths large with from 20 to 25 rather short tentacles, scattered without any very definite arrangement in rows. The most distal tentacles are farther from the extremity of the proboscis than is usual in the genus.

Gonosome.—Several sporosacs, as many as 8 or 10, are developed from the body just above or just below the proximal tentacles. Female sporosacs large with few large ova; male not so large as the female but still of good size. Very often the majority of the sporosacs develop on the one side of the hydranth and give it a distorted appearance.

Distribution.—Friday Harbor (Fraser).

Genus SYNCORYNE

Trophosome.—Colony unbranched or slightly and irregularly branched; perisarc well developed; tentacles strongly capitate.

Gonosome.—Gonophores usually few in number; medusae with four rudimentary tentacles.

KEY TO SPECIES

A. Colony much branched, perisarc irregularly annulated.......
....*S. eximia*
B. Colony unbranched or slightly branched, perisarc not annulated.................................*S. mirabilis*

Syncoryne eximia (Allman)
Plate 3, Fig. 11

Coryne eximia ALLMAN, Ann. and Mag. N.H., (3) IV, 1859, p. 141.
Syncoryne eximia HINCKS, Br. Hyd. Zooph., 1868, p. 50.
FRASER, West Coast Hyd., 1911, p. 21.

Trophosome.—Colonies in tangled masses, each much branched, but with the branches almost entirely on the one side of the stem, or, when the large branches are branched, on the one side of each branch; annulations irregular, often represented by a wrinkling only, seldom present to the same extent on the ultimate branches; hydranths rather elongated with 20 to 30 tentacles.

Gonosome.—Medusa buds, each on a short pedicel, scattered among the tentacles, from the body of the hydranth.

Distribution.—Juneau, Alaska (Nutting); Pacific Grove, Cal. (Torrey); near Lime point, San Francisco bay, 18 fathoms; Petersburgh, Alaska.

Syncoryne mirabilis (Agassiz)
Plate 3, Fig. 12

Coryne mirabilis AGASSIZ, Contr. Nat. Hist. U.S., IV, 1862, p. 185.
Coryne rosaria A. AGASSIZ, Ill. Cat., 1865, p. 176.
Syncoryne mirabilis TORREY, Hyd. of the Pacific coast, 1902, p. 31.
FRASER, West Coast Hyd., 1911, p. 21.
Vancouver I. Hyd., 1914, p. 114.

Trophosome.—Colony unbranched or slightly and irregularly branched; stem and branches similar in size; perisarc smooth; hydranth body stout for its length; tentacles 15 or more, stout, strongly capitate.

Gonosome.—Gonophores borne among or below the proximal tentacles; medusae become sexually mature before liberation.

Distribution.—San Francisco (A. Agassiz); Santa Barbara (Fewkes); San Francisco bay (Torrey); gulf of Georgia (A. Agassiz); Puget sound (Calkins); Bare island (Hartlaub); San Juan archipelago, Queen Charlotte islands, Departure bay, Five Finger islands, Pylades channel, Gabriola pass, Porlier pass, off Matia island, Puget sound, western Skidegate narrows, low tide (Fraser); at the entrance to San Francisco bay, point Lobos, and in the middle section of the bay, Angel island to west Berkeley; Eel river bar and off Heceta head, Oregon; Channel islands, Puget sound; Juneau, Garforth island, Muir inlet, Alaska; low tide to 10 fathoms.

Family Atractylidae

Trophosome.—Perisarc well developed; hydranths with a basal whorl of filiform tentacles and conical proboscis.

Gonosome.—Gonophores producing sporosacs or free medusae.

KEY TO GENERA

A.   Gonophores producing sporosacs
    a. Blastostyle forming a branched spadix........*Atractyloides*
    b. Blastostyle unbranched
        a. Sporosacs permanently surrounded by perisarc..*Bimeria*
        b. Sporosacs not permanently surrounded by perisarc....
        ....*Garveia*
B.   Gonophores producing free medusae
    a. Medusae with clusters of tentacles..........*Bougainvillia*
    b. Medusae with all tentacles arranged singly....*Perigonimus*

Genus ATRACTYLOIDES

This genus was established by Fewkes to accommodate a new species found off Santa Barbara, Cal., but no definite generic characteristics are given.   There is nothing in the description of the species that gives any indication of a reason for separating it from *Garveia* or *Bimeria* unless it is the complex development of the "spadix" in the sporosac.

Atractyloides formosa Fewkes

Plate 3, Fig. 13

Atractyloides formosa FEWKES, New invertebrates from the coast of California, 1889, p. 5.
FRASER, West Coast Hyd., 1911, p. 22.

Trophosome.—"Stem solitary, erect, brownish, with masses of attached algae on their external surfaces; the distal ends funnel-shaped; attached to a creeping stolon. Each polypite (hydranth) projects from a cup-shaped hydrotheca. Hydranth with single circle of tentacles; mouth and intratentacular region of whitish colour. Hydrothecal base annulated" (Fewkes).

Gonosome.—"The sporosacs arise from the base of attachment on solitary, erect stems. Each male capsule has a central axis (spadix) which has a green and yellow colour; near its distal end the spadix is enlarged into a disk-shaped structure, and about midway in its length there arise lateral branches from which originate the spermatic masses. The proximal part of the spadix is connected with the inner wall of the male capsule by a network of fibres. At the distal end of the same organ, the walls of the spadix and those of the capsule are similarly united. Spermatic elements are formed inside the sporosac and are developed from the external wall of the spadix, probably making their way out through an opening in the distal end of the sporosac.

"The female *Atractyloides* was not observed" (Fewkes).

Distribution.—Off the coast of Santa Barbara, Cal. (Fewkes).

This species has not been reported since Fewkes first described it. His description and figures are used.

Genus **BIMERIA**

Trophosome.—Colony branched, invested with a conspicuous perisarc; hydranths fusiform; perisarc covering the base of the tentacles.

Gonosome.—Sporosacs covered with perisarc throughout the whole period of development, arising from the stems by means of very short pedicels.

KEY TO SPECIES

A. Stem not fascicled or but slightly fascicled
    a. Stem and branches slender
        a. Pedicels not annulated or wrinkled.........*B. pusilla*
        b. Pedicels much annulated or wrinkled.......*B. tenella*
    b. Stem and branches stout, in dense tufts....*B. franciscana*
B. Stem strongly fascicled
    a. Branches slender, but little wrinkled..........*B. gracilis*
    b. Branches stout, wrinkled throughout..........*B. robusta*

Bimeria franciscana Torrey
Plate 3, Fig. 14

Bimeria franciscana TORREY, Hyd. of the Pacific coast, 1902, p. 28.
FRASER, West Coast Hyd., 1911, p. 22.

Trophosome.—Stems stout, up to 7 cm. in length, occurring in dense clusters; primary branches at quite regular intervals, sometimes given off in the same plane but more commonly in all planes; secondary branches given off from the distal side of the primary branches, either themselves terminating in hydranths or giving off smaller branchlets that bear hydranths; perisarc of the main stem smooth, but usually there is a slight annulation or wrinkling in the proximal portion; the branches are annulated or wrinkled throughout. The cup-like expansion, surrounding the body of the hydranth may be transversely wrinkled; hydranths with 14-16 tentacles.

Gonosome.—Sporosacs, almost sessile, or with short pedicels, appear on the secondary branches, or on the hydranth-bearing branchlets, in irregular clusters, principally in connection with the larger primary branches, and as the ultimate branches are closely placed in this region, the sporosacs are rather crowded.

Distribution.—San Francisco bay (Torrey); various scattered locations in each of the three sections of San Francisco bay. Low tide to 7 fathoms.

Bimeria gracilis Clark
Plate 4, Fig. 15

Bimeria gracilis CLARK, Hyd. of the Pacific coast, 1876, p. 252.
TORREY, Hyd. of San Diego, 1904, p. 6.
FRASER, Vancouver Island Hyd., 1914, p. 115.
Hyd. of the west coast of V.I., 1935, p.143.
Hyd. Distr. in the vicinity of Q.C.I., 1936,
p. 123.

Trophosome.—Stem fascicled, growing from a creeping stolon, reaching a height of 35 mm.; branches rather short and delicate, seldom spreading but taking nearly the same direction as the stem; perisarc mostly smooth but slightly ringed or wrinkled at the base of each pedicel; hydranths with 10-11 tentacles.

Gonosome.—Gonophores borne on the branches, singly or in pairs; sporosacs oval; pedicel short, almost suppressed.

Distribution.—San Diego (Clark); Nanoose bay, off Clarke Rock, off West rocks, Dodds narrows, Gabriola pass, Gabriola reefs, Ruxton passage; off Sydney inlet; entrance to southwest arm and west of Horn island in Tasoo harbour, western end of Houston Stewart channel, off Rose harbour (Fraser); in scattered locations in the three sections of San Francisco bay and outside the Golden Gate; off Heceta head light, Ore.; off cape James, Hope island. 3 to 67 fathoms.

? Bimeria pusilla Fraser
Plate 4, Fig. 16

Bimeria pusilla Fraser, Some new and some previously unreported Hyd., 1925, p. 168.

Trophosome.—Colony small, straggling, less than 5 mm. high; from one to three main branches that are similar in size and appearance to the main stem; secondary branches vary in length and are irregularly placed; these may bear hydranths or may divide again to form pedicels for hydranths; the angle that the small branches make with the larger, and the larger make with the stem, is very variable; the perisarc is thin, nowhere annulated or wrinkled, but there is a slight tendency to waviness that prevents complete smoothness; this waviness appears in the small branches as well as in the larger and the stem; hydranths with 12-14 tentacles.

Gonosome.—Unknown.

Distribution.—Lime point, San Francisco bay (Fraser).

? Bimeria robusta Torrey
Plate 4, Fig. 17

Bimeria robusta Torrey, Hyd. of the Pacific coast, 1902, p. 29.
Fraser, West coast Hyd., 1911, p. 22.
Vancouver Island Hyd., 1914, p. 115.

Trophosome.—Colony large, up to 13 cm., forming a rather dense tuft; main branches large, irregular; stem and main branches fascicled, the tubes forming the fascicle twisting much around one another, never passing in a straight course for any great distance; the secondary branches are short and either give rise to hydrothecae themselves or divide again to form hydranth pedicels. Each branch or branchlet forms a slight angle with its support and as the branchlets are numerous, this gives a much crowded effect. Perisarc wavy or wrinkled in almost all parts but never annulated; hydranths with 11 or 12 tentacles.

Gonosome.—Unknown.

Distribution.—On float at Ferry landing, San Pedro, Cal. (Torrey); San Juan archipelago (Fraser).

? **Bimeria tenella** Fraser

Plate 4, Fig. 18

Bimeria tenella FRASER, Some new and some previously unreported hyd., 1925, p. 168.

Trophosome.—Stem simple or with a slight tendency to fasciculation, reaching a height of 15 mm.; stem and branches slender, the branches making a wide angle with the stem or branches; the pedicels attached to the stem in general much longer than those attached to the branches; the main stem and branches mainly smooth, but there may be a wrinkling or even an annulation for a short distance; pedicels annulated extensively at the base or more rarely annulated or wrinkled throughout their whole length; the perisarc surrounding the base of the hydranth is heavy and somewhat rough and wrinkled; hydranth with 10 tentacles.

Gonosome.—Unknown.

Distribution.—Several locations near the entrance to San Francisco bay in the neighbourhood of Southampton light, Angel island and Alcatraz island. Depth, 9-22 fathoms (Fraser).

Genus **GARVEIA**

Trophosome.—Colony branched or unbranched; perisarc conspicuous, reaching well up on the fusiform hydranth.

Gonosome.—Gonophores borne on distinct branch-like pedicels. The sporosacs may be temporarily enclosed in thin perisarc, but this bursts off in the later stages so that the perisarc is confined to the pedicels, where it usually ends in a cup-like expansion.

KEY TO SPECIES

A. Zooids growing directly from a stolon............*G. formosa*
B. Stem simple, pedicel slender...............*G. groenlandica*
C. Stem fascicled
   a. Stem strongly fascicled, annulated throughout.. *G. annulata*
   b. Stem loosely fascicled, slightly or not at all annulated.....
   ....*G. nutans*

Garveia annulata Nutting

Plate 5, Fig. 19

Garveia annulata NUTTING, Hyd. of Harriman Exp. 1901, p. 166.
Bimeria annulata TORREY, Hyd. of the Pacific coast, 1902, p. 28.

3

Garveia annulata FRASER, West Coast Hyd., 1911, p. 22.
                         Vancouver Island Hyd., 1914, p. 117.
                         Hyd. of west coast of V.I., 1935, p. 143.
                         Hyd. Distr. in the vicinity of Q.C.I.,
                         1936, p. 122.

Trophosome.—Stem fascicled, but the various parts that go to make it up are more lightly held together than is usual in a polysiphonic stem. Sometimes branches that give rise to two or more pedicels come off the main stem, but more commonly the pedicels arise directly from it. The whole colony may reach a height of 50 mm. The pedicels vary much in length and in the amount of their annulation; some of them are regularly annulated throughout while others are wrinkled rather than annulated. There is little regularity in their arrangement or in their attitude as compared with the stem. The hydranth is large, fusiform, with the base enclosed in perisarc, which extends to the base of the tentacles; they are about 16 in number.

Gonosome.—The gonophores are borne on the stem and on the stolon; the pedicel is branch-like, covered with thick perisarc, which extends to end in a shallow cup-like expansion below the sporosac, the remains of a ruptured sporosac envelope. The sporosac is large, oval or nearly globular.

Distribution.—Yakutat and Sitka, Alaska (Nutting); Santa Catalina island and San Francisco, Cal. (Torrey); Port Renfrew, Ucluelet, Queen Charlotte islands; Clayquot sound, Northumberland channel, Dodds narrows, Gabriola pass, Whaleboat passage, off Shaw island, San Juan channel, very abundant in the neighbourhood of Friday Harbor, particularly off point Richardson; Estevan point, Bajo reef, bar off Indian village, Esperanza inlet; north of Marble island, west of Horn island, Tasoo harbour, entrance to Flamingo harbour, western entrance to Houston Stewart channel, off Rose harbour, west of cape St. James, Massett harbour (Fraser); distributed through all of San Francisco bay and out through the Golden Gate, Dillon's beach, Catalina island, Cal.; off Heceta head light, Ore.; Nawhitti bar, off Hope island; Sitka and Yakutat, Alaska. Depth, 3-64 fathoms.

With hydranths and pedicels light orange and main stem and sporosacs a deeper orange, this is quite a conspicuous species.

Garveia formosa (Fewkes)
Plate 5, Fig. 20

Perigonimus formosus FEWKES, New Invert. of Cal. Coast, 1889, p.6.
Garveia formosa TORREY, Hyd. of the Pacific coast, 1902, p. 8.
FRASER, West Coast Hyd., 1911, p. 23.

Trophosome.—Zooids arising individually from a stolon that is but slightly branched and does not form a definite reticulum; pedicels varying in length up to 4 mm.; perisarc smooth or but very slightly wrinkled, not enlarged at the base of the hydranth but is thinned out from the inside to accommodate the hydranth; tentacles 10-16.

Gonosome.—Gonophores borne on short pedicels growing singly from the stolon. In the early stages there is a thick envelope of perisarc.

Distribution.—Santa Cruz, Cal. (Fewkes); near Southampton light in San Francisco bay. Depth, 7 fathoms.

Garveia groenlandica Levinsen
Plate 5, Fig. 21

Garveia groenlandica LEVINSEN, Meduser, Ctenophorer og Hydroider
fra Groenlands Vestkyst, 1893,
p. 155.
FRASER, Vancouver Island Hyd., 1914, p. 117.
Hyd. from the west coast of V.I.,
1935, p. 143.
Hyd. Distr. in the vicinity of Q.C.I.,
1936, p. 123.

Trophosome.—Stems arising from a reticulated stolon, simple, usually unbranched but sometimes a main stem gives rise to two or three branches, each similar to the unbranched stem; much more delicate than *G. annulata*; pedicels vary in length from 2 to 6 mm.; perisarc wrinkled or somewhat irregularly annulated, passing over the body of the hydranth to the base of the tentacles; tentacles 10 in number.

Gonosome.—Gonophores borne on the stolon; sporosacs oval or globular, borne on distinct pedicels, nearly as long as the vertical diameter of the sporosacs. The sporosac is first covered with a thin perisarc which is like a thin sack, but this soon ruptures, leaving a somewhat wrinkled or flapped portion to form the cup at the base of the sporosac. The peduncle of the sporosac is very short so that the base of the sporosac is but a short distance above the cup-like prominence of the perisarc of the pedicel.

Distribution.—Dodds narrows, Gabriola pass, Gabriola reefs, Whaleboat passage, Swiftsure shoal, off Matia island, off O'Neale island, Port Townshend; off Sydney inlet, V.I.; off Rose harbour, off Massett harbour, Q.C.I.; off Bull harbour and north of Hope island, off Klashwan point, Graham island; off McArthur reef, Sumner strait, Alaska.   13 to 50 fathoms.

Garveia nutans Wright
Plate 5, Fig. 22

Garveia nutans WRIGHT, Edin. Phil. Jour., 1859, p. 109.
                   HINCKS, Br. Hyd. Zooph., 1868, p. 102.
                   FRASER, West Coast Hyd., 1911, p. 23.

Trophosome.—Stem loosely fascicled, irregularly branched; branches vary much in length, slightly annulated if at all but more commonly with the perisarc somewhat wrinkled; pedicels often bent abruptly near the hydranth; the terminal portion of the perisarc expands somewhat to form a funnel-shaped cup; hydranth with about 10 tentacles.

Gonosome.—Gonophores borne on short branch-like pedicels, the perisarc on the pedicel ending in a cup-like expansion, some distance before the gonophore is reached; gonophores large, oval or orbicular, male and female similar in size and shape.

Distribution.—Berg inlet, Glacier bay, Alaska (Nutting).

Genus BOUGAINVILLIA

Trophosome.—Colonies fascicled, much branched; perisarc well developed on stems and branches; hydranths fusiform, proboscis dome-shaped or conical.

Gonosome.—Gonophores supported on short pedicels; medusae with four radial canals and four clusters of tentacles; tentacle bulb ocellate.

KEY TO SPECIES

A.   Branches twining around the stem; gonophores in clusters....
     ....*B. glorietta*
B.   Branches loosely arranged; gonophores single.....*B. mertensi*

Bougainvillia glorietta Torrey

Plate 6, Fig. 23

Bougainvillia glorietta TORREY, Hyd. of San Diego, 1904, p. 7.

FRASER, West Coast Hyd., 1911, p. 23.

Vancouver Island Hyd., 1914, p. 119.

Trophosome.—Stem fascicled, branched, reaching a height of 20 to 30 cm.; branches with a tendency to twine around the stem; perisarc smooth or wavy, but without annulations; tentacles about 20 in number, in an irregular whorl.

Gonosome.—Gonophores borne on the branches, in clusters of 3 or sometimes only 2. Medusae with four pairs of tentacles.

Distribution.—San Diego, Cal. (Torrey); near the entrance of Hammond bay (Fraser).

Bougainvillia mertensi Agassiz

Plate 6, Fig. 24

Bougainvillia mertensi AGASSIZ, Cont. Nat. Hist. U.S., IV, 1862, p. 344.

A. AGASSIZ, N.A. Acalephae, 1865, p. 152.

FRASER, West Coast Hyd., 1911, p. 24.

Vancouver Island Hyd., 1914, p. 119.

Trophosome.—Main stem thick, fascicled, reaching a height of 10 cm.; branching loose, larger branches fascicled; hydranths on rather short pedicels, far apart on the distal branches but not so much so on the proximal; perisarc smooth or wavy, slightly annulated on the pedicels.

Gonosome.—Gonophores borne singly on short branches, that in some cases are without hydranths. There may be more than one gonophore on a branch but they are not in pairs or clusters. The gonophore is provided with a short pedicel.

Distribution.—Bering sea, San Francisco (A. Agassiz); Oakland (Torrey); gulf of Georgia (Agassiz); Nanoose bay, Dodds narrows, Gabriola pass, Griffin bay, Upright channel, Friday Harbor (Fraser); Brothers island light, Goat island and Alameda channel, in San Francisco bay. 3 to 10 fathoms.

Genus PERIGONIMUS

Trophosome.—Colony unbranched or slightly branched; hydranths clavate with conical or dome-shaped proboscis.

Gonosome.—Gonophores bearing medusae that when liberated have 2-4 marginal tentacles arranged singly; no ocelli.

KEY TO SPECIES

A. Perisarcal tube of the same diameter throughout....*P. repens*
B. Perisarcal tube tapering from apex to base........*P. serpens*

Perigonimus repens (Wright)
Plate 6, Fig. 25

Eudendrium pusillum WRIGHT, Proc. Roy. Phys. Soc. Edin., 1857, p. 231.

Atractylis repens WRIGHT, *ibid.*, 1858, p. 450.

Perigonimus repens ALLMAN, Ann. and Mag. N.H., (3), VIII, 1864, p. 365.

FRASER, West Coast Hyd., 1911, p. 24.

Vancouver Island Hyd., 1914, p. 120.

Hyd. Distr. in vicinity of Q.C.I., 1936, p. 123.

Trophosome.—Colonies small; stems unbranched, forked, or slightly and irregularly branched, arising from a reticular stolon; perisarc well developed but apparently fitting very loosely over the coenosarc; the tube of the same diameter throughout; it is not expanded distally but is large enough that the hydranth may retract within it.   Tentacles about 10 in number, standing out rather stiffly from the hydranth body.

Gonosome.—Gonophores borne on pedicels growing from the stolon or the stem.   In the latter case there may be one or more present on the same stem.   Medusae with two developed and two rudimentary tentacles at the time of liberation.

Distribution.—Townshend harbour (Calkins); Sausolito, Cal., between tides (Torrey); Departure bay, off Lasqueti island, off north end of Gabriola island, Northumberland channel; Houston Stewart channel, off Rose harbour, west of Rose spit (Fraser); Quarry point and Goat island in San Francisco bay; Point Gardiner buoy, Admiralty island and Symonds point, Lynn canal, Alaska. 9-30 fathoms.

Perigonimus serpens Allman
Plate 6, Fig. 26

Perigonimus serpens ALLMAN, Ann. and Mag. N.H., (3), XI, 1863, p. 10.

HINCKS, British Hyd. Zooph., 1868, p. 95.

FRASER, Hyd. from the Q.C.I., 1935, p. 505.

Trophosome.—Zooids growing singly from a filiform stolon, perisarc well developed, the tube tapering slightly from apex to base; hydranth oval or clavate, with 8-12 extensile tentacles.

Gonosome.—Gonophores growing from the stolon on short pedicels; medusae with two tentacles at the time of liberation.

Distribution.—Rennell sound, 60-70 fathoms; Tasoo harbour, 7-14 fathoms; west of Rose spit, 16 fathoms (Fraser).

Family **Eudendridae**

Trophosome.—Colony usually branching; perisarc well developed; proboscis trumpet-shaped but with much freedom of movement; tentacles all filiform in a single whorl.

Gonosome.—Gonophores producing fixed sporosacs; male and female gonophores dissimilar; male gonophores in whorls, female in clusters.

Genus **EUDENDRIUM**

The only genus of the family *Eudendridae.*

KEY TO SPECIES

A. Stems and branches strongly annulated
   a. Stems coarse
      a. Rings very distinct and close together..*E. californicum*
      b. Rings less distinct and farther apart.....*E. vaginatum*
   b. Stems slender............................. *E. insigne*
B. Stems and branches with few annulations
   a. Stems fascicled
      a. Main stem, primary and secondary branches fascicled
              ....*E. rameum*
      b. Main stem only, fascicled..............*E. ramosum*
   b. Stems simple
      a. Hydranths entirely aborted on pedicels that bear gonophores.......................*E. capillare*
      b. Hydranths not aborted.................*E. tenellum*
   c. No definite stems........................*E. irregulare*

Eudendrium californicum Torrey
Plate 6, Fig. 27

Eudendrium californicum TORREY, Hyd. of the Pacific coast, 1902, p. 32.

FRASER, West Coast Hyd., 1911, p. 24.
Vancouver Island Hyd., 1914,
p. 121.
Hyd. as a food supply, 1933,
p. 261.
Hyd. from west coast of V.I.,
1935, p. 144.

Trophosome.—Stems stout, simple, in clusters from an encrusting plate-like hydrorhiza; branches stiff and short as compared with the length of the main stem, given off in all planes, each making a wide angle with the stem but distally turning in nearly the same direction as the stem; hydranths large with about 20 tentacles; perisarc on stem, branches, and pedicels, very distinctly annulated with narrow annulations, extending over the body of the hydranth to the base of the tentacles.

Gonosome.—Female gonophores monothalamic, crowded on the body of the hydranth immediately proximal to the tentacles; male gonophores bithalamic, in two or three whorls, just proximal to the tentacles.

Distribution.—San Francisco bay, Tomales bay, Pacific Grove, Cal. (Torrey); Santa Cruz, Monterey bay (Clark); Port Renfrew, Ucluelet; Northumberland channel; in the stomach of *Melanitta perspicillata*, Bay city, Ore.; Estevan point, near Maquinna point, off Nootka island, Bajo reef (Fraser); several locations in the upper and middle sections of San Francisco bay and out through the Golden Gate, Aumentos rock, Pinos light, Pacific Grove, Cal.; off Heceta head, Ore.; off cape James, Hope island. 2-63 fathoms.

Eudendrium capillare Alder
Plate 7, Fig. 28

Eudendrium capillare ALDER, Ann. and Mag. N.H., (2), XVIII,
1856, p. 355.
FRASER, West Coast Hyd., 1911, p. 24.
Vancouver Island Hyd., 1914,
p. 122.

Trophosome.—Colony small, not over 15 mm. in height, usually branched, annulations at the base of the branches and pedicels.

Gonosome.—Female gonophores borne on aborted hydranths which are supported by pedicels springing from branches or from the stolon. They form a very noticeable cluster. Male gonophores dithalamic, in whorls, borne similarly to the female gonophores.

Distribution.—San Juan archipelago; off Matia island, Friday Harbor (Fraser); off Yaquina light, Ore.; off McArthur reef, Sumner strait, Point Gardiner buoy, Admiralty island, off Symonds point, Lynn canal, all in Alaska.   9 to 40 fathoms.

Eudendrium insigne Hincks
Plate 7, Fig. 29

Eudendrium insigne Hincks, Ann. and Mag. N.H., (3), VIII, 1861,
p. 159.
Br. Hyd. Zooph., 1868, p. 86.
Fraser, Vancouver Island Hyd., 1914, p. 122.
Hyd. from west coast of V.I., 1935,
p. 144.

Trophosome.—Stolons forming an irregular network from which the small colonies spring; stem simple, slender, with few branches given off irregularly.   A branch is usually about the same size as the main stem, hence the branching appears dichotomous.   Hydranths with 20-25 tentacles.   There is a characteristic furrow passing horizontally around the body of the hydranth nearly halfway to the base of the tentacles from which the gonophores spring. The stolon is not annulated but the stem, branches, and pedicels are annulated throughout.

Gonosome.—Female gonophores globular, on short stalks; male gonophores in a whorl, 2 chambered, each chamber oval.

Distribution.—Clayuquot sound, Pylades channel; bar off Indian village, Esperanza inlet (Fraser).

Eudendrium irregulare Fraser
Plate 7, Fig. 30

Eudendrium irregulare Fraser, A new Hydractinia and other west coast Hyd., 1922, p. 97.

Trophosome.—Stolons irregular, straggling over such other hydroids as *Lafoea gracillima*, not forming a regular network but, in places, adhering to form a loose fascicle; there are no definite stems as there are in most of the species of this genus.   Most commonly the pedicels grow singly from the stolon; they are more or less sinuous and they vary greatly in length, the longest about 4 mm. Probably because the stolons grow over more or less erect hydroids, these pedicels pass out in all directions and at various angles. Neither the stolons nor the pedicels are annulated, although the perisarc may be slightly and irregularly wrinkled. Hydranth with few tentacles, 8 to 10.

Gonosome.—Unknown.

Distribution.—Northumberland channel, 15 fathoms (Fraser).

Eudendrium rameum (Pallas)

Plate 7, Fig. 31

Tubularia ramea Pallas, Elench. Zooph., 1766, p. 83.
Eudendrium rameum Torrey, Hyd. of the Pacific coast, 1902, p. 33.
              Fraser, West Coast Hyd., 1911, p. 25.
              Vancouver Island Hyd., 1914, p. 122.
              Hyd. Distr. in vicinity of Q.C.I., 1936, p. 123.

Trophosome.—Stem large, fascicled, much and irregularly branched; large branches may also be fascicled; small branches give rise to the pedicels for the hydranths, these usually passing out from the distal side of the branches; hydranths with 24-25 tentacles. Perisarc on main stem and on large branches smooth or at most wavy, small branches with a few annulations or wrinkles at the base, pedicels annulated throughout.

Gonosome.—Female gonophores oval, borne in a cluster from the hydranth body below the base of the tentacles. Male gonophores similarly placed, forming a whorl, 2 or 3-chambered.

Distribution.—Swiftsure shoal, off Matia island, off Waldron island, Friday Harbor; in Houston Stewart channel, in the western portion of the channel and off Rose harbour (Fraser); on float, Ferry landing, San Pedro, Cal. (Torrey); San Pedro, Bluff point, San Francisco bay, Cal.; off Symonds point, Lynn canal, Alaska. 9 to 30 fathoms.

Eudendrium ramosum (Linnaeus)

Plate 7, Fig. 32

Tubularia ramosa Linneaus, Syst. Nat., 1758, p. 804.
Eudendrium ramosum Torrey, Hyd. of the Pacific coast, 1902, p. 34.
              Fraser, West Coast Hyd., 1911, p. 25.

Trophosome.—Stem slightly fascicled, much and irregularly branched; height 15 cm.; hydranth pedicels usually vertically placed on the pinnately arranged branches; annulations at the base of the branches and pedicels.

Gonosome.—Both male and female gonophores borne at the base of the hydranths or some distance down the pedicels; hydranths normal or reduced in size; male gonophores with 2 or 3 chambers, usually few in number.

Distribution.—Pacific Grove, San Diego, Cal. (Torrey); San Pedro, Pacific Grove, off Bare island light and Brothers island light in San Francisco bay, Cal.   2-10 fathoms.

Eudendrium tenellum Allman
Plate 8, Fig. 33

Eudendrium tenellum ALLMAN, Gulf Stream Hyd., 1877, p. 8.
FRASER, Vancouver Island Hyd., 1914, p. 123.
Hyd. Distr. in vicinity of Q.C.I.,
1936, p. 123.

Trophosome.—Colony very small, seldom reaching a height of 1 cm., slender, growing from a stolon which forms a loose network over worm tubes, *etc.*; stem unbranched or with one or more branches, irregularly arranged and forming a wide angle with the stem; hydranth with about 20 tentacles; perisarc smooth for the most part, usually 2 or 3 annulations at the base of the stem and branches and often 2 or 3 together along the stem or branch, apparently at no definite distance from the base.

Gonosome.—Gonophores arranged around the base of the hydranth body, the female oval, the male two-chambered, the chambers nearly globular; the gonophores may spring from the stolon or from the branches.

Distribution.—Near Clarke rock, north of Gabriola island, Northumberland channel; in tow net, near surface, at ebb tide, western portion of Houston Stewart channel (Fraser); off Ball point, San Diego, Cal.; Sumner strait, off Shingle island, McArthur reef, Alaska.   Surface to 240 fathoms.

Eudendrium vaginatum Allman
Plate 8, Fig. 34

Eudendrium vaginatum ALLMAN, Ann. and Mag. N.H., (3), II, 1863,
p. 10.
NUTTING, Harriman Hyd., 1901, p. 167.

FRASER, West Coast Hyd., 1911, p. 25.

Vancouver Island Hyd., 1914, p. 124.

Hyd. as a food supply, 1933, p. 260.

Hyd. from west coast of V.I., 1935, p. 144.

Hyd. Distr. in vicinity of Q.C.I., 1936, p. 123.

Trophosome.—Colonies growing in small clusters may reach a height of 40 mm.; main stem much larger than the branches, the latter usually short and apparently loosely connected with the stem; tentacles fewer than 20; perisarc annulated throughout, the annulations not very close together and not very distinct at times; the perisarc passes up on the body of the hydranth, forming a cup-shaped portion that may be quite smooth or may be somewhat wrinkled.

Gonosome.—Male gonophores 2-chambered, in a whorl about the base of the hydranth that is not aborted; female gonophores in dense clusters around the body of hydranths usually devoid of tentacles.

Distribution.—Akutan pass, Alaska (Clark); Sitka harbour and Yakutat, Alaska (Nutting); Swiftsure shoal; in the stomach of *Clangula hyemalis*, St. Paul's island, Bering sea; on Danger rocks near eastern entrance to Houston Stewart channel (Fraser); off Heceta head and Yaquina point, Ore.; Yakutat and Whale island, Alaska. Low tide to 75 fathoms.

Family Hydractinidae

Trophosome.—Colony formed of distinct nutritive and generative zooids, growing from a common basal coenosarc, which secretes a calcareous crust, commonly beset with spines; other kinds of zooids may also be present. Hydranths with a row of filiform tentacles; proboscis conical or clavate.

Gonosome.—Gonophores giving rise to fixed sporosacs on special generative zooids.

Genus HYDRACTINIA

The only genus of the family *Hydractinidae.*

KEY TO SPECIES

A. Spines numerous, jagged; sporosacs and ova numerous......
....*H. aggregata*

B. Spines represented by irregular protuberances, ova 1 or 2 . . . . .
. . . .*H. californica*
C. Spines smooth
    a. Spines short, slightly curved; sporosacs in the distal half
        of the hydrocaulus; one ovum . . . .*H. laevispina*
    b. Spines long, straight; sporosacs near middle of hydrocaulus;
        one ovum in the sporosac . . . . *H. milleri*

Hydractinia aggregata Fraser
Plate 8, Fig. 35

Hydractinia aggregata FRASER, West Coast Hyd., 1911, p. 25.
                        Vancouver Island Hyd., 1914, p. 124.
                        Hyd. Distr. in vicinity of Q.C.I., 1936, p. 123.

Trophosome.—Nutritive zooids at many different stages of growth may be found in the same colony. They grow from a basal coenosarc that is well supplied with jagged spines; these may be conical like those of *H. echinata*, they may be much blunter, more in the nature of columns, or these may be joined to form a ridge of some length. The zooids in the contracted state may be entirely below the tips of these spines. The number of tentacles increases during development until the number in the adult reaches 20-24.

Gonosome.—Sporosacs begin to develop on the generative zooids when they are very small, at which time they, *i.e.*, the zooids, have a greater number of tentacles, 10-12, than when the sporosacs are fully developed, as then 3 or 4 seems to be the regular number. The generative as well as the nutritive zooids have mouths. The sporosacs appear some distance below the tentacles, about one-fourth the distance to the base of the zooid. The ova are large and numerous. The male sporosacs are oval, not nearly so large as the female.

Other zooids.—None observed.

Distribution.—Departure bay, San Juan archipelago; common on gastropod shells all along the strait of Georgia near Departure bay, Friday Harbor; entrance to southwest arm, Tasoo harbour, west of Rose spit (Fraser). 10-25 fathoms.

Hydractinia californica Torrey

Hydractinia californica TORREY, Hyd. of San Diego, 1904, p. 9.
FRASER, West Coast Hyd., 1911, p. 27.

Trophosome.—"Sterile hydranths 2 to 2.5 mm. long in extension, with 6 to 10 tentacles, usually in two recognizable whorls; proboscis domed to conical; spines .5 to .9 mm. long, often with truncated tops and irregular protuberances; with about 10 longitudinal dental ridges" (Torrey).

Gonosome.—"Sporosacs, with one or two eggs in the female, borne in clusters of 2 to 10 or more about half way from the base of the blastostyle. Latter with 5 to 10 knob-like clusters of nematocysts, representing tentacles, 1 to 1.3 mm. long" (Torrey).

Distribution.—San Diego, Cal., 50 fathoms (Torrey).

I have not seen this species. The description is Torrey's. No figures are given.

Hydractinia laevispina Fraser
Plate 8, Fig. 26

Hydractinia laevispina FRASER, A new Hydractinia and other west coast hyd., 1922, p. 97.

Trophosome.—In a colony, the nutritive zooids appear in successive stages of growth, the largest reaching a length of 2.5 mm. Even in the largest zooids the tentacles are few in number as compared with *H. aggregata*, eight being the common number. There are few spines, smooth, rather blunt, slightly curved. They are about 0.5 mm. in length.

Gonosome.—Sporosacs begin to develop on the generative zooids when they are small; on these zooids there are three or four tentacles that persist although they remain small; a mouth is present. The sporosacs, commonly four in number, appear at about one-third of the distance from the tentacles to the base. The female sporosacs are small, with one ovum in each; the male considerably larger; both spherical.

Other zooids.—Scattered about the outer portion of the colony are numerous long, slender, tentacular filaments or "tentaculozooids". They are only one-fourth the diameter of the extended nutritive zooids and may be twice as long. They are well provided with nematocysts but have neither mouth nor tentacles.

Distribution.—On barnacles in 7-10 fathoms, at the western end of Gabriola pass (Fraser).

Hydractinia milleri Torrey
Plate 8, Fig. 37

Hydractinia milleri TORREY, Hyd. of the Pacific coast, 1902, p. 24.
FRASER, West Coast Hyd., 1911, p. 27.
Vancouver Island Hyd., 1914, p. 125.
Hyd. from west coast of V.I., 1935, p. 144.

Trophosome.—Colony growing from a basal coenosarc, which is provided with long smooth spines; nutritive zooids robust when mature, reaching a height of 5 mm. The clavate proboscis is capable of great extension, but it may be contracted until it is little more than a knob. There are 12-20 tentacles in an irregular whorl which, under certain conditions of contraction, shows an arrangement into sets of four.

Gonosome.—Generative zooids shorter and more slender than the nutritive; tentacles fewer in number but never entirely lacking; sporosacs borne low on the hydrocaulus, about midway between the tentacles and the base of the zooid; they are not numerous, four being the greatest number noticed on one zooid. Female sporosacs are small with usually but one ovum; male sporosacs larger.

Other zooids.—"Spiral zooids at the edge of the colony, about as long as the sterile hydranths, but much more slender, the whole structure resembling a very long tentacle" (Torrey).

Distribution.—Between tides, Tomales bay, San Francisco, Cal. (Torrey); Port Renfrew (Fraser).

Family **Pennaridae**

Trophosome.—Colony branched; hydranths with a proximal whorl of long filiform tentacles around the body of the hydranth and several capitate tentacles on the proboscis, these usually in a series of whorls.

Gonosome.—Gonophores producing free medusae with four radial canals and four rudimentary tentacles.

Genus **PENNARIA**

Trophosome.—Colony large, much branched, often with a distinct pinnate or twice pinnate arrangement; hydranths with a large proboscis, very noticeable when extended, well supplied with whorls of capitate tentacles.

Gonosome.—Gonophores borne on the hydranth body just distal to the proximal whorl of tentacles; medusae large, often mature, when liberated. They may even liberate the sex products before being set free from the hydranth.

Pennaria tiarella McCrady
Plate 9, Fig. 38

Pennaria tiarella McCrady, Gymno. of Charleston Har., 1857, p. 51.

Fraser, Beaufort Hyd., 1912, p. 355.

Trophosome.—Colony large, sometimes reaching the height of 15 cm.; branching twice pinnate. A varying number of annulations, never many, occur on the main stem above the origin of the branch and on the branches above their origin. The hydranths are large, narrowing distinctly to form the proboscis; there are 10 or 12 filiform tentacles and a varying number of capitate tentacles which are usually arranged in 4 or 5 quite regular whorls in the fully developed hydranth. Often a hydranth bud appears growing directly from the wall of the main stem or branch.

Gonosome.—Gonophores few in number; when there are more than one on a hydranth at the same time, they seldom are at the same stage of development. The medusae are oval or ovate; rudimentary tentacles, radial canals and the manubrium are all well developed, and the sexual products may be dehisced, before the medusae are set free.

Distribution.—Marin island, San Francisco bay, 4½ fathoms.

Family Corymorphidae

Trophosome.—Zooids solitary, large; hydranths with a proximal and a distal set of filiform tentacles.

Gonosome.—Gonophores producing free medusae with four radial canals, and three of the four tentacles aborted or very much reduced.

Genus CORYMORPHA

Trophosome.—Pedicel with perisarc represented by a thin pellicle; tubular, fleshy processes growing from the pedicel near the base; hydranth abruptly distinct from the pedicel; proximal tentacles longer than the distal; distal set in several contiguous rows.

Gonosome.—Gonophores borne on branched pedicels between the two sets of tentacles.

KEY TO SPECIES

A. Proximal tentacles about 40 . . . . *C. carnea*
B. Proximal tentacles 18-30 . . . . *C. palma*

Corymorpha carnea (Clark)

Rhizonema carnea CLARK, Alaskan Hyd., 1876, p. 233.
Corymorpha carnea TORREY, Hyd. of the Pacific coast, 1902, p. 9.
FRASER, West Coast Hyd., 1911, p. 27.

Trophosome.—Stem 68 mm. long, of nearly uniform size for two-thirds of the distance towards the base, then increasing rapidly to quite a sharp point, like an acorn. The processes for attachment are developed in large numbers from the basal surface of the enlarged portion of the stem. Proximal tentacles about 40, slender, reaching a little beyond the proboscis; distal tentacles numerous, delicate. Length of head 13 mm., basal portion 21 mm. (taken from Clark's original description).

Gonosome.—Gonophores borne on branched peduncles, between the two sets of tentacles (taken from Clark's original description).

Distribution.—Norton sound, Alaska (Clark).

Apparently only the two specimens of this species recorded by Clark have yet been seen. With his description no figures were given.

Corymorpha palma Torrey
Plate 9, Fig. 39

Corymorpha palma TORREY, Hyd. of the Pacific coast, 1902, p. 37.
Hyd. of San Diego region, 1904, p. 9.
FRASER, West Coast Hyd., 1911, p. 27.

Trophosome.—Zooids, up to 14 cm., rooted in sand by a dense tangle of filamentous processes, each process arising from a papilla-like structure; stem somewhat bulbous at the base, the remainder, except for a small distal portion, tapering somewhat towards the base of the hydranth body, not pendulous; proximal tentacles 18-30, in one whorl; distal tentacles very numerous, in several irregular whorls, delicate.

Gonosome.—Gonophores borne on peduncles attached to the body of the hydranth between the proximal and the distal tentacles, medusoid, with four radial canals and a manubrium, which may be several times as long as the bell.

Distribution.—San Pedro, San Diego, Cal., between tides (Torrey); in numerous collections from San Pedro, Cal.

4

Family Tubularidae

Trophosome.—Colony branched irregularly or unbranched; hydranths with a proximal and a distal set of filiform tentacles.

Gonosome.—Gonophores, borne on the hydranth between two sets of tentacles, give rise to actinulae.

Genus TUBULARIA

Trophosome.—Colony unbranched or branched irregularly; hydranths large; proximal set of tentacles longer than the distal set.

Gonosome.—Gonophores in clusters, attached by means of stalked peduncles to the body of the hydranth just distal to the proximal tentacles; female gonophores producing actinulae.

KEY TO SPECIES

A. Colony branched
    a. Perisarc extensively annulated................*T. larynx*
    b. Perisarc not extensively annulated.............*T. crocea*
B. Colony unbranched
    a. Distal tentacles more numerous than proximal
        a. Proximal tentacles 40, distal tentacles slightly more numerous.........................*T. indivisa*
        b. Proximal tentacles 30-32; distal 50-60.....*T. borealis*
        c. Proximal tentacles 20; distal 30...........*T. aurea*
    b. Distal tentacles less numerous than proximal
        a. Proximal tentacles not more than 26; distal not more than 18.............................*T. marina*
        b. Proximal tentacles 40-50; distal 20......*T. harrimani*

Tubularia aurea Fraser
Plate 9, Fig. 40

Tubularia aurea FRASER, Hyd. from Queen Charlotte Is., 1936, p. 504.

Trophosome.—Individual zooids growing from a slightly annulated stolon to a height of 3 cm.; the pedicel has a heavy perisarc, almost to the distal end, without annulations. Hydranths with about 20 proximal and 30 distal tentacles.

Gonosome.—Gonophores large, in loose racemes, with few to each cluster. The terminal processes are quite large.

Colour.—Stolon and pedicels, a rich, golden yellow; hydranth white.

Distribution.—Danger rocks at the eastern end of Houston Stewart channel, low tide; Flamingo harbour, on gastropod shells, attached to *Macrocystis.*

Tubularia borealis Clark

Tubularia borealis CLARK, Alaskan Hyd., 1876, p. 231.

FRASER, West Coast Hyd., 1911, p. 27.

Trophosome.—"Hydrocaulus simple, erect, straight, annulated or twisted at the base, smooth, not forming a collar-like expansion below the hydranth, light straw colour. Hydranths large, drooping; proximal tentacles 30-32 in number, arranged in a single verticil, when expanded, forming a circle with a diameter of 20 to 24 mm.; distal tentacles 50 to 60, arranged in a double row so closely as to form a single verticil" (Clark).

Gonosome.—"Gonophores oviform, with four crest-shaped, laterally compressed, tentaculiform processes, arranged in pendant racemes of 6 to 7 in number, which, when mature, are twice the length of the hydranth, and bear from 30 to 70 sporosacs each" (Clark).

Distribution.—Hagmeister island, Alaska (Clark).

No one but Clark has reported this species and he gave no figures.

Tubularia crocea (Agassiz)
Plate 9, Fig. 41

Parypha crocea AGASSIZ, Contr. Nat. Hist. U.S., IV, 1862, p. 249.

Tubularia crocea FRASER, West Coast Hyd., 1911, p. 28.

Vancouver Island Hyd., 1914, p. 127.

Hyd. Distr. in vicinity of Q.C.I., 1936, p. 123.

Trophosome.—Colony growing in immense tufts which make a tangled mass below, but separated into long pedicels which reach out above the mass; branching very irregular; stems slightly and irregularly annulated, somewhat swollen just below the hydranth; proximal and distal tentacles nearly equal in number, 20-24.

Gonosome.—Gonophores growing in rather long racemes, which, however, seldom hang below the tentacles; each is provided with four short tentacular processes but they may be so short as to be almost suppressed.

Distribution.—San Francisco (A. Agassiz); San Francisco bay, San Pedro, San Diego (Torrey); Port Simpson, B.C., gulf of Alaska;

Gabriola pass, Porlier pass, Friday Harbor; western portion of Houston Stewart channel (Fraser); San Diego, Coronado, San Pedro, Codiga bay, numerous locations in all three sections of San Francisco bay, Cal. Low tide to 18 fathoms.

Tubularia harrimani Nutting
Plate 10, Fig. 42

Tubularia harrimani Nutting, Hyd. of the Harriman Exp., 1901, p. 168.

Fraser, West Coast Hyd., 1911, p. 28.

Vancouver Island Hyd., 1914, p. 127.

Trophosome.—Stem usually unbranched, reaching a height of 40-50 mm., there are few annulations but these are usually very distinct; stem slender at the base, rapidly increasing in size proximally and more slowly distally toward the hydranth; proximal tentacles much more numerous than the distal, there being 40-50 proximal and only about 20 distal.

Gonosome.—Gonophores borne on several long, densely-crowded racemes; each gonophore is provided with 3 or 4 short tentacles that may be almost as long as the gonophore itself.

Distribution.—Orca, Prince William sound, Alaska (Nutting); Port Renfrew, off Matia island, off Brown island, Friday Harbor (Fraser).

Tubularia indivisa Linnaeus
Plate 10, Fig. 43

Tubularia indivisa Linnaeus, Syst. Nat., 1767, p. 1301.

Fraser, West Coast Hyd., 1911, p. 28.

Vancouver Island Hyd., 1914, p. 128.

Trophosome.—Stems seldom branched, growing in clusters; much the largest species of these here reported (Hincks says it may reach a height of 12 inches) and the perisarc is heavier than in any of the other species, with the possible exception of *T. aurea*; the stem may be twisted at the base but there are no distinct annulations; the proximal tentacles are long, slender, and numerous, up to 40, but the distal are even more numerous.

Gonosome.—Gonophores in racemes but not such long ones as in *T. harrimani*; gonophores devoid of tentacular processes.

Distribution.—St. Michael's, Norton sound, Alaska (Clark); Alert bay, off Waldron island, off Brown island, Friday Harbor (Fraser).

Tubularia larynx Ellis and Solander
Plate 10, Fig. 44

Tubularia larynx ELLIS and SOLANDER, Nat. Hist. Zooph., 1786,
p. 31.
CALKINS, Some Hyd. of Puget Sd., 1899, p. 335.
FRASER, West Coast Hyd., 1911, p. 28.
Vancouver Island Hyd., 1914, p. 128.

Trophosome.—Colony consisting of stems much branched and somewhat tangled at the base, usually extensively annulated but the nature of the annulations vary; they may be deeply cut as in some of the *Eudendridae*, but more commonly the furrows as well as the ridges are rounded to give the surface a wavy appearance. Tentacles approximately the same number in the proximal and distal sets, about 20.

Gonosome.—Gonophores arranged in clusters that are denser and more compact than in other species described. The tentacular processes are not extensively developed.

Distribution.—Port Townshend (Calkins); Nanoose bay, Gabriola pass, off point Richardson and at other points near Friday Harbor, Samish bay (Fraser); off cape James, Hope island, 30 fathoms.

Tubularia marina Torrey
Plate 11, Fig. 45

Tubularia marina TORREY, Hyd. of the Pacific coast, 1902, p. 46.
FRASER, West Coast Hyd., 1911, p. 28.

Trophosome.—Stems clustered, reaching a height of 5 cm., unbranched, of much the same thickness throughout, but commonly increasing slightly distally; irregularly annulated or wavy, more especially in the proximal portion, often entirely smooth; distal extremity not so distinctly bulbous as in *T. crocea*; hydranths with a maximum of 26 proximal and 18 distal tentacles.

Gonosome.—Gonophores in long racemose clusters, tending to be pendulous; male gonophores broadly ovate, with apical processes slightly developed; female gonophores more narrowly ovate, with long, stout, apical processes, each with a bulbous base.

Distribution.—Trinidad, San Francisco, Pacific Grove, Cal. (Torrey); outside of Golden Gate and in several locations in all three sections of San Francisco bay; Channel rocks, Puget sound. 4-20 fathoms.

A. Agassiz mentions a species under the name, *Thamnocnidia tubularoides* (Ill. Cat., 1865, p. 196), but the description is so meagre, without any figures, that it is not possible to place it definitely. There is much variation in *Tubularia crocea* in the San Francisco bay region. Some with small heads might correspond to *Parypha microcephala* Agassiz, and the very large ones might correspond to *Thamnocnidia tubularoides* Agassiz, but no constant difference is observable, with the exception of the size of the hydranth, from the typical *T. crocea*, and that does not seem sufficient to make distinct species. As it is, the young *T. crocea* is difficult to distinguish at times from that of *T. marina*.

Family **Hybocodonidae**

Trophosome.—"Colony unbranched. Stem with a distinct chitinous perisarc, and rooted by a true hydrorhiza. Hydranths large, with a proximal and distal set of filiform tentacles" (Nutting).

Gonosome.—"Gonophores producing free medusae" (Nutting).

Genus **HYBOCODON**

Trophosome.—"Stem with distinct, deeply annulated expansion just below the hydranth. Hydranth with a proximal whorl and two distinct but closely approximated distal whorls of filiform tentacles" (Nutting).

Gonosome.—"Gonophores attached directly to the hydranth body without the intervention of peduncles and developing into free medusae. The medusae are deeply campanulate with four radial canals and short proboscis" (Nutting).

Hybocodon prolifer Agassiz
Plate 11, Fig. 46

Hybocodon prolifer AGASSIZ, Contr. Nat. Hist. U.S., IV, 1862, p.243.
    BIGELOW, Proc. U.S. Nat. Mus., 1913, p. 6.
    FRASER, Vancouver Island Hyd., 1914, p. 130.

Trophosome.—"Hydrocaulus unbranched, longitudinally striped, owing to the coenosarcal canals showing through; perisarc suddenly enlarging near the hydranth, where a number of collar-like swollen rings appear, the uppermost being the largest. Hydranth much like that of *Tubularia* but with two distinctly separate whorls of tentacles around the proboscis, each whorl being composed of about 16 tentacles, the lower being twice as long as the upper" (Nutting).

Gonosome.—"Gonophores adnate to the hydranth body just above the base whorl of tentacles, producing free medusae with four radial canals and 5 superficial meridional orange-colored bands when fully mature. The single tentacle is greatly enlarged and near its base a number of medusae in various stages of development are attached and these again in the same manner may bear other groups of medusae" (Nutting).

Distribution.—Medusae only. Dutch Harbor, Alaska (Bigelow); Departure bay (Fraser).

Sub-order CALYPTOBLASTEA

Hydroids with hydranths protected with hydrothecae and gonophores protected by gonangia or similar structures.

KEY TO FAMILIES

A. Hydrothecae free from the stem or branches (with exception of some of the hydrothecae in the fascicled *Lafoeidae*).

    a. Hydrotheca campanulate, with diaphragm but without operculum

        a. Hydranth without preoral cavity.....*Campanularidae*

        b. Hydranth with preoral cavity.........*Bonneviellidae*

    b. Hydrotheca tubular

        a. Hydrotheca with diaphragm

            1. Hydrotheca with operculum of converging segments...................*Campanulinidae*

            2. Hydrotheca without operculum........*Hebellidae*

        b. Hydrotheca without diaphragm or operculum........
        ....*Lafoeidae*

    c. Hydrotheca reduced to saucer-shaped hydrophore........
    ....*Halecidae*

B. Hydrotheca sessile and more or less adnate to the stem or branch

    a. Stem and branches without nematophores

        a. Hydrotheca margin entire; no operculum .*Synthecidae*

        b. Hydrotheca margin usually with teeth; operculum present............................*Sertularidae*

    b. Nematophores on stem or branches or both..*Plumularidae*

Family **Bonneviellidae**

Trophosome.—Hydranths with a perforated membrane (veloid) stretching from the tentacle bases above the real oral surface, thus forming a preoral cavity.

Gonosome.—Gonophores producing sporosacs or free medusae.

Genus **BONNEVIELLA**

Trophosome.—"Hydranth with a single row of tentacles connected by a veloid, forming a cavity, preoral, lined with ectoderm. The branched colony arising from a rhizocaulon" (Nutting).

Gonosome.—"Gonangia scattered over the stem or in groups on rootstock. Gonophores sessile. Colonies sexually distinct" (Nutting).

KEY TO SPECIES

A. Hydrotheca more than 1 cm. deep; slightly campanulate, margin smooth...........................*B. superba*
B. Hydrotheca less than 5 mm. deep; tubular or slightly urceolate; margin sinuous.............................*B. regia*

Bonneviella regia (Nutting)
Plate 12, Fig. 47

Campanularia regia NUTTING, Hyd. of Harriman Exp., 1901, p. 172.
FRASER, West Coast Hyd., 1911, p. 32.
Vancouver Island Hyd., 1914, p. 138.
Bonneviella regia, NUTTING, Am. Hyd., III, 1915, p. 95.

Trophosome.—Stems unbranched serving as pedicels, arising from a stolon; pedicels often shorter than the hydrothecae, without annulations or with one or two; hydrotheca large, reaching a length of 3.5 mm. and a breadth of 1.25 mm., almost tubular in some cases but more urceolate in others; margin slightly everted and slightly sinuous, the sinuosities being wide but very shallow; reduplication of the margin often takes place; hydranth with 18-20 tentacles.

Gonosome.—Gonangium large, but not so, relatively to the hydrotheca; it is about 2/3 the length and 2/3 the diameter of the hydrotheca; it is deeply corrugated, each corrugation being provided with a keel.

Distribution.—Orca, Prince William sound, Alaska (Nutting); off Matia island, off Sucia island, very plentiful in some Friday Harbor material (Fraser); off Klashwan point, Graham island, 30 fathoms, Bering sea, 43 fathoms.

Bonneviella superba Nutting

Plate 12, Fig. 48

Bonneviella superba NUTTING, Am. Hyd., III, 1915, p. 96.

Trophosome.—"Colony consisting of a tangled mass of partly adherent tubes which form an axis or stem from which single pedicels arise. This pseudo-stem is interwoven with a mass of other hydroids, mostly sertularians. Pedicels strong, stiff, attaining a length of 2.5 cm., and a diameter of over 1 mm. They are perfectly smooth, for the most part, but are constricted near their origin and just below the hydrotheca. Hydrothecae enormous, in one case attaining a length of 1.7 cm., probably the largest hydrotheca known. The diameter near the margin is 6 mm. The general shape is deep campanulate rather than tubular, diminishing gradually below until it passes into the pedicel, and flaring at the margin above. Margin perfectly smooth. There is no real diaphragm, although there appears to be one, as the bottom of the hydranth is free from the hydrothecal floor. There is no chitinous shelf, however.

The hydranths are very large, with a single circle of smooth tentacles. There is no proboscis. The surface, which would, without histological investigation, be taken for the oral disk, being almost perfectly flat. A longitudinal section of hydranth shows that this apparent oral surface is in reality the 'veloid' of Broch, and that it covers a distinct preoral chamber of considerably greater size than that of *B. grandis*, which is, as in that species, lined with ectoderm. Below this and perhaps surrounding its conical lower part, is the gastric cavity, lined with convoluted endoderm" (Nutting).

Gonosome.—"The gonangia are in an aggregated cluster of cylindrical bodies growing from a tangled mass, much as in the case of several species of *Lafoea*. Individual gonangia attain a length of 6 mm. and a diameter of 1.5 mm. They are rudely annulated throughout, there being 7 broad corrugations in the one described. There is a broad neck, almost as broad as the rest of the gonangium, and an abruptly truncated end. The structure is supported on a short pedicel. The gonangial contents have generally discharged or are partially disintegrated, so that a satisfactory investigation can hardly be made" (Nutting).

Distribution.—Bering sea, 283 fathoms (Nutting).

Family **Campanularidae**

Trophosome.—Hydrotheca campanulate, never sessile, never adnate to, or immersed in, the stem or branches; diaphragm always present; hydranth with trumpet-shaped proboscis.

Gonosome.—Gonophores producing sporosacs or free medusae; the medusae when produced usually have lithocysts and have the gonads along the course of the radial canals.

KEY TO GENERA

A. Gonophores producing sporosacs in which the planulae are developed
  a. Sporosacs remain within the gonangia during development
  ....*Campanularia*
  b. Sporosacs are extruded into a sac at the summit of the gonangium, in which sac the planulae are developed..
  ....*Gonothyraea*
B. Gonophores producing medusoids without mouth or digestive cavity.....................................*Eucopella*
C. Gonophores producing free medusae
  a. Medusae globular, with four tentacles at liberation..*Clytia*
  b. Medusae flatter, with 16 or more tentacles at liberation..
  ....*Obelia*

Genus **CAMPANULARIA**

Trophosome.—As in the family.
Gonosome.—Gonophores producing sporosacs from which planulae develop within the gonangia.

KEY TO SPECIES

A. Stem fascicled
  a. Hydranth pedicels in whorls...............*C. verticillata*
  b. Hydranth pedicels given off singly.........*C. gelatinosa*
B. Stem branched but not fascicled.................*C. exigua*
C. Stem simple, unbranched or slightly branched
  a. Hydrothecal margin entire
    a. Hydrotheca campanulate.................*C. integra*
    b. Hydrotheca nearly tubular................*C. ritteri*
  b. Hydrothecal margin toothed
    a. Hydrotheca with vertical lines

1. Lines very distinct throughout the whole length of the hydrotheca
     i. Gonangium annulated . . . . *C. hincksi*
    ii. Gonangium smooth . . . . *C. groenlandica*
2. Lines distinct towards margin only . . . . *C. speciosa*
b. Hydrotheca without vertical lines
  1. Teeth blunt
      i. Teeth tapering from base to tip
         I. Hydrotheca large . . . . *C. gigantea*
        II. Hydrotheca relatively small
             x. Teeth 5 or 6 . . . . *C. raridentata*
            xx. Teeth about 10 . . . . . . *C. volubilis*
           xxx. Teeth 12 to 18 . . . . *C. urceolata*
     ii. Teeth as broad at base as at tip . . . *C. castellata*
  2. Teeth acute
      i. Teeth deep, hydrotheca tapering from margin to base . . . . *C. denticulata*
     ii. Teeth shallow, hydrotheca suddenly narrowing at base . . . . *C. fusiformis*

? Campanularia castellata Fraser

Plate 12, Fig. 49

Campanularia castellata FRASER, Some new and some previously unreported hydroids, 1925, p. 170.

Trophosome.—Stem unbranched, forming the hydranth pedicel, from 0.8 mm. to 4.0 mm. in length, terminated below the hydrotheca in a ball-shaped joint, but otherwise only slightly annulated, if at all; in some cases there is a slight waviness near each extremity; stolon not annulated; hydrotheca large, 0.6 to 1.0 mm. in length and 0.4 to 0.6 mm. in greatest width, tapering but slightly from the margin to near the base, the base nearly hemispherical; the 12-14 teeth are deep, the same width throughout or slightly broader at the tip, which is just noticeably rounded; the hydrothecal wall is uniform in thickness, without lines or ridges; the space between the diaphragm and the base is shallow.

Gonosome.—Unknown.

Distribution.—Near Alcatraz island, San Francisco bay, growing on *Sertularia desmoides*. 10-17 fathoms (Fraser).

Campanularia denticulata Clark

Plate 12, Fig. 50

Campanularia denticulata CLARK, Alaskan Hyd., 1876, p. 213.

FRASER, West Coast Hyd., 1911, p. 29.

Vancouver Island Hyd., 1914, p. 132.

NUTTING, Am. Hyd., III, 1915, p. 52.

Trophosome.—Stems usually unbranched but occasionally one branch also giving rise to a hydrotheca, appears, growing from a stolon which is not annulated to any great extent; the stem or pedicel of the hydranth varies much in length and the amount of annulation; usually there are several annulations at the base and fewer at the distal end below the hydrotheca. Hydrotheca deeply campanulate, tapering very gradually from margin to base; teeth sharp, deep, about 15 in number.

Gonosome.—Gonangia arising from the stem or from the stolon, with short, annulated pedicels; oblong-ovate; the distal portion does not narrow to form a collar, the aperture being the full width of the gonangium.

Distribution.—Port Etches, Alaska (Clark); Orca, Alaska (Nutting); San Pedro (Torrey); Departure bay, San Juan archipelago; near Round island, in Dodds narrows, off Matia island (Fraser).

Campanularia exigua (Sars)

Plate 12, Fig. 51

Laomedea exigua SARS, Middelhavet's Littoral Fauna, 1857, p. 50.

Campanularia exigua CALKINS, Hyd. from Puget Sound, 1899, p. 353.

FRASER, West Coast Hyd., 1911, p. 30.

Vancouver Island Hyd., 1914, p. 134.

NUTTING, Am. Hyd., III, 1915, p. 48.

Trophosome.—"Stem very delicate, slightly flexuous, giving off at each bend simple pedicels, ringed at the base and upper extremity (the intermediate space being smooth) which support the pedicels; height about one-quarter inch; hydrothecae very small, regularly funnel-shaped, with an even rim" (Hincks).

Gonosome.—"Gonothecae axillary, elongate, smooth, somewhat fusiform" (Hincks).

Distribution.—Port Townshend (Calkins).

Campanularia fusiformis Clark
Plate 12, Fig. 52

Campanularia fusiformis CLARK, Hyd. of Pacific Coast, 1876, p.254.
FRASER, West Coast Hyd., 1911, p. 134.
Vancouver Island Hyd., 1914, p. 134.
NUTTING, Am. Hyd., III, 1915, p. 52.

Trophosome.—"Hydrocaulus simple, creeping, bearing the pedicels at regular intervals; pedicels of variable length, usually two or three times the length of the hydrotheca, never more than six times their length, with a more or less wavy outline. Hydrotheca small, deeply campanulate, tapering at the base, rim ornamented with about 12 stout, shallow, acute teeth, a single distinct annulation at the base" (Clark).

Gonosome.—"Gonothecae small, fusiform, constricted at both ends, sessile; aperture small, terminal" (Clark).

Distribution.—Point Reyes peninsula, Dillon's beach, Cal., between tides (Torrey); Vancouver island (Clark); Monterey bay (Nutting).

Campanularia gelatinosa (Pallas)
Plate 13, Fig. 53

Sertularia gelatinosa PALLAS, Elench. Zooph., 1766, p. 116.
Laomedea pacifica AGASSIZ, Ill. Cat., 1865, p. 194.
Campanularia pacifica TORREY, Hyd. of the Pacific coast, 1902, p. 53.
FRASER, West Coast Hyd., 1911, p. 32.
Obelia gelatinosa FRASER, West Coast Hyd., 1911, p. 39.
Campanularia gelatinosa FRASER, Vancouver Island Hyd., 1914, p. 135.
Obelaria gelatinosa NUTTING, Am. Hyd., III, 1915, p. 88.
Campanularia gelatinosa FRASER, Hyd. Distr. in vicinity of Q.C.I., 1936, p. 123.

Trophosome.—Stem fascicled, growing in clusters, reaching a height of 20 to 25 cm., larger branches are also fascicled; in the fascicled portions the perisarc is thickened and dark in colour, but in the smaller branches and their ramifications, it is whitish transparent. As the small branches divide somewhat dichotomously, a large number of hydranth pedicels appear close together and these in their whiteness give the gelatinous appearance when in the water, to which evidently the specific name is due. The branches have usually 3-5 annulations at the base and the larger branches from which they spring have a similar number above their point of origin. The hydranth pedicels are slender, varying much in length; the shorter ones are annulated throughout but the longer ones may have a smooth portion towards the centre. The hydrothecae are deeply campanulate, tapering quite gradually from margin to base; margin with about 10 teeth, each provided with two cusps.

Gonosome.—Gonangia elongated-oval with distinct neck and tapering base; pedicel short, annulated.

Distribution.—Discovery bay, Wash. (Calkins); California (Jäderholm); gulf of Georgia (A. Agassiz); San Francisco bay (Torrey); San Juan archipelago; Nanoose bay, near Clarke rock, Northumberland channel, Dodds narrows, Whaleboat passage; off Matia island, Friday Harbor, off Brown island, Upright channel, off O'Neale island, off Waldron island, San Juan channel, Port Townshend; in Houston Stewart channel at the western end and off Rose harbour, Massett harbour (Fraser); middle section of San Francisco bay, mainly near Oakland and Berkeley, off Bull harbour, west end of Houston Stewart channel, off Klashwan point, Q.C.I. Low tide to 80 fathoms.

? Campanularia gigantea Hincks
Plate 13, Fig. 54

Campanularia gigantea HINCKS, Ann. and Mag. N.H., (3), XVIII, 1866, p. 297.
Br. Hyd. Zooph., 1868, p. 174.
FRASER, Hyd. from Queen Charlotte Is., 1936, p. 507.

Trophosome.—Zooids growing singly from a stolon, or, in some cases, one or more other zooids are attached to the main pedicel as branches, in which case the branch is similar to the main zooid; pedicels long and slender, the whole colony reaching a height of 15 mm.; annulated at the base and below the hydrotheca; hydrotheca very large, over a mm. in length, campanulate, the length nearly twice the greatest breadth, with 12-14 regularly rounded teeth, the indentations being similar in curvature to the teeth.

Gonosome.—Unknown.

Distribution.—9½ miles south of Marble island, Q.C.I., 190 fathoms (Fraser).

Campanularia groenlandica Levinsen
Plate 13, Fig. 55

Campanularia groenlandica LEVINSEN, Meduser, Ctenophorer og
Hydroider, 1893, p. 26.
FRASER, West Coast Hyd., 1911, p. 31.
Vancouver Island Hyd., 1914,
p. 136.
NUTTING, Am. Hyd. III, 1915, p. 36.
FRASER, Hyd. from west coast of V.I.,
1935, p. 144.
Hyd. Distr. in vicinity of
Q.C.I., 1936, p. 123.

Trophosome.—Stem unbranched, forming the hydranth pedicel,
annulated throughout but the stolon is not annulated; hydrotheca
large, somewhat urceolate, although at times the sides are almost
straight, the base being hemispherical; the 10-12 teeth are quite
deep, rounded or squared off at the tip; the wall of the hydrotheca
is somewhat ribbed; the line so formed showing quite distinctly,
running from the spaces between the teeth, the full length of the
hydrotheca.

Gonosome.—Gonangium large, bottle-shaped, with a long neck,
attached to the stolon with a very short pedicel; surface smooth; ova
very few, large.

Distribution.—Berg inlet, Glacier bay, Alaska (Nutting); Puget
sound (Nutting); Port Renfrew; off cape Edenshaw, Swiftsure shoal,
off Lasqueti island, Northumberland channel, off Matia island, off
Sucia island, off O'Neale island, Friday Harbor; off Sydney inlet;
Houston Stewart channel, Masset harbour (Fraser); San Diego,
Cal.; off Klashwan point, off Massett sound, off cape James, Hope
island; off McArthur reef, Sumner strait, Gardiner buoy, Admiralty
island, Alaska.   29-50 fathoms.

Campanularia hincksi Alder
Plate 13, Fig. 56

Campanularia hincksi ALDER, Trans. Tynes. Field Club, 1857, p. 37·
TORREY, Hyd. of the Pacific coast, 1902, p. 53.
FRASER, West Coast Hyd., 1911, p. 31.
NUTTING, Am. Hyd., III, 1915, p. 37.

Trophosome.—Stem unbranched, forming the pedicel for the
hydranth, long, slender, annulated below the hydrotheca and at the

base; hydrothecae deep, nearly tubular, with lines running from the margin almost to the base; margin with square-topped teeth.

Gonosome.—Gonangia borne on the stolon, ovoid, truncate, corrugated; pedicel short, not annulated.

Distribution.—San Diego, Cal., shallow water (Torrey); San Diego, Mare island in San Francisco bay, 7 fathoms.

Campanularia integra MacGillivray
Plate 13, Fig. 57

Campanularia integra MacGillivray, Ann. and Mag., N.H., (2),
IX, 1842, p. 465.

Clark, Alaskan Hyd., 1876, p. 215.

Fraser, West Coast Hyd., 1911, p. 31.

Vancouver Island Hyd.,1914, p. 137.

Nutting, Am. Hyd., III, 1915, p. 33.

Fraser, Hyd. from west coast of V.I., 1935,
p. 144.

Hyd. Distr. in vicinity of Q.C.I.,
1936, p. 123.

Trophosome.—Stem unbranched forming the pedicel for the hydranth, arising from a stoloniferous network; the pedicels are long and slender, varying much in the amount of the annulation; there is always one clear-cut annulation, accompanied by two or three others less deep, at the base of the hydrotheca; hydrotheca rather small, tapering gradually from margin to base; margin entire.

Gonosome.—Gonangium large, deeply corrugated, each corrugation with a distinct keel; attached to the stolon by a short, annulated pedicel.

Distribution.—Lituya bay and Shumagin islands, Alaska(Clark); Bering sea (Jäderholm); off Southern California, 36 fathoms (Nutting); Port Wilson, Port Townshend and Bremerton, Wash. (Calkins); San Juan archipelago; off Round island, in Dodds narrows, Gabriola pass, off Waldron island; gulf of Alaska; off Clayoquot sound, off Sydney inlet, Estevan point, near Maquinna point, Bajo reef: entrance to Flamingo harbour, Houston Stewart channel at the western end and off Rose harbour (Fraser); west of Nawhitti bar, off Sadler point, off Virago sound and off Massett, Q.C.I.; off McArthur reef, Sumner strait, off Shingle island, Sumner strait, Alaska.   Low tide to 240 fathoms.

A species, *Campanularia occidentalis*, was mentioned by Fewkes in "New Invertebrata from the coast of California, 1899, p. 4", as occurring at Santa Barbara, but the description is so meagre that it is impossible to place it.

? Campanularia raridentata Alder
Plate 13, Fig. 58

Campanularia raridentata ALDER, Ann. and Mag. N.H., (3), IX, 1862, p. 315.

FRASER, West Coast Hyd., 1911, p. 32.

Vancouver Island Hyd., 1914, p. 138.

NUTTING, Am. Hyd., III, 1915, p. 39.

FRASER, Hyd. Distr. in vicinity of Q.C.I., 1936, p. 123.

Trophosome.—Stem unbranched, serving as the pedicel of the hydranth, arising from a stolon which at this point has a distinct elevation, somewhat bulbous in appearance; pedicel annulated at the base and below the hydrotheca and sometimes more or less throughout; hydrotheca long and narrow, tapering but slightly from margin to base; teeth usually 5 in number, deep and rounded at the tip.

Gonosome.—Unknown.

Distribution.—Departure bay, Queen Charlotte islands, Rose spit, Northumberland channel, Friday Harbor; Houston Stewart channel, off Rose harbour, off Massett inlet (Fraser); lower section of San Francisco bay; western portion of Goose island halibut bank. 1 to 35 fathoms.

A species, which he named *Laomedea rigida*, was reported by A. Agassiz in the "Illustrated Catalogue", 1865, p. 93, but the description is not complete enough for identification. It is as follows:

"This species is remarkable for its peculiar mode of growth. At first glance it would readily be mistaken for a species of *Dynamena*, so regular is the succession of the hydrae along the stem, and also on account of the absence of branches. The sterile and reproductive hydrae are found on the sides of the main stem, attached by a very short pedicel, and alternately so regularly on each side that its campanularian nature is noticed only after a careful examination. The sterile hydrae resemble those of *Laomedea amphora* while the reproductive calyces are identical in shape with those of *Obelia com-*

5

missuralis. The main stems of a cluster are crowded together and attain a height of three or four inches."

Nutting suggests that it may be the same as *Clytia bakeri*, later described by Torrey.

Campanularia ritteri Nutting
Plate 13, Fig. 59

Campanularia ritteri NUTTING, Harriman Hyd., 1901, p. 171.
FRASER, West Coast Hyd., 1911, p. 33.
NUTTING, Am. Hyd., III, 1915, p. 35.

Trophosome.—"Usually consisting of unbranched pedicels growing directly from a creeping rootstock, which is not regularly annulated. Pedicels long and slender, often two or three times the length of the hydrotheca, with two or three annulations at the proximal end and a spherical annulation at the distal end. Otherwise they are usually without distinct annulation. The hydrothecae are long for this genus, tubular, their sides being nearly parallel and with a round, perfectly smooth rim" (Nutting).

Gonosome.—Gonangia large, about three times as long as wide, coarsely and regularly annulated, with annulation nearly horizontal. Pedicel very short, borne on the rootstock and not annulated. Gonangial contents numerous, developing ova closely packed around the blastostyle" (Nutting).

Distribution.—Juneau, Alaska, and off the California coast (Nutting).

Campanularia speciosa Clark
Plate 13, Fig. 60

Campanularia speciosa CLARK, Alaskan Hyd., 1876, p. 171.
FRASER, West Coast Hyd., 1911, p. 33.
Vancouver Island Hyd., 1914, p. 139.
NUTTING, Am. Hyd., III, 1915, p. 48.
FRASER, Hyd. Distr. in vicinity of Q.C.I., 1936, p. 123.

Trophosome.—Stem unbranched, serving as pedicel, arising from an annulated stolon; the pedicels are also annulated throughout; they are short as compared with the length of the hydrotheca and may be shorter than it is. Hydrotheca large, reaching a height of 2 mm., urceolate; margin rather undulated than toothed as the teeth are low and rounded; teeth 10-12 in number; from the teeth, lines run down the wall of the hydrotheca for some distance but they cannot be traced more than one-third of the distance to the base.

Gonosome.—Gonangium shaped like an inverted cone, except that the sides are somewhat curved; the height and greatest width are nearly equal; the pedicel is very short.

Distribution.—Shumagin islands, Alaska (Clark); Orca, Yukon harbour, Big Koniushi, Shumagin islands, Yakutat bay, Alaska, 6-20 fathoms (Nutting); Friday Harbor off Massett, Gabriola pass, off Matia island; Houston Stewart channel at the west end and near Rose harbour; western Goose island halibut grounds, off Frederick island, off cape James on Hope island.   Surface to 50 fathoms.

Campanularia urceolata Clark
Plate 13, Fig. 61

Campanularia urceolata CLARK, Alaska Hyd., 1876, p. 215.

FRASER, West Coast Hyd., 1911, p. 33.

Vancouver Island Hyd., 1914, p. 140.

NUTTING, Am. Hyd., III, 1915, p. 40.

FRASER, Hyd. from west coast of V.I., 1935, p. 144.

Hyd. Distr. in vicinity of Q.C.I., 1936, p. 123.

Trophosome.—Stems unbranched, serving as pedicels, arising from a stolon that is usually smooth when attached to a surface (usually of other hydroids) but strongly annulated when it is free; pedicels varying much in length, usually annulated or wavy throughout; hydrothecae very variable in size, shape, and nature of margin; they may be tubular, urceolate, or turgid towards the base; the margin has 12-18 teeth, usually shallow and blunt; reduplication of the margin often takes place.

Gonosome.—Female gonangia bottle-shaped, with long neck; surface wavy or slightly corrugated; attached to stolon with very short, annulated pedicels; there appears to be but one sporosac present, containing 7 or 8 ova.   The male gonangia are short, not much deeper than wide and without the bottle-neck; the sinuosities are more deeply cut and that gives the gonangium an irregular shape.

Distribution.—Port Etches, Lituya bay, Alaska (Clark); Santa Cruz, Cal. (Clark); Yakutat, Alaska (Nutting); San Francisco bay,

Pacific Grove, Tomales bay, Cal. (Torrey); almost everywhere where dredging has been done around the Queen Charlotte islands and Vancouver island (Fraser); Dillon's beach, Santa Cruz, point Reyes peninsula, Pacific Grove, outside the Golden Gate, numerous locations in the middle section of San Francisco bay and some in the adjacent portions of the upper and lower sections; Newport and off Heceta head, Ore.; many more locations in the Vancouver island and the Queen Charlotte region; Sitka, Yakutat, and Kadiak, Alaska. 1 to 67 fathoms.

Campanularia verticillata (Linnaeus)
Plate 14, Fig. 62

Sertularia verticillata LINNAEUS, Syst. Nat., 1767, p. 1311.
Campanularia verticillata FRASER, West Coast Hyd., 1911, p. 33.
Vancouver Island Hyd., 1914,
p. 140.
NUTTING, Am. Hyd., III, 1915, p. 29.
FRASER, Hyd. from west coast of V.I.,
1935, p. 145.
Hyd. Distr. in vicinity of Q.C.I.,
1936, p. 123.

Trophosome.—Main stem fascicled throughout, ending like a stump; main branches also fascicled; hydranths arranged in regular whorls, with rather long pedicels, annulated or wavy throughout. Hydrothecae large, broad for their length, slightly more expanded towards the margin; margin with 12-14 low blunt teeth.

Gonosome.—Gonangium somewhat fusiform except that the distal end is prolonged into a neck, sessile on the stem; often occurring in groups around the stem, although not forming a whorl; ova large.

Distribution.—Port Etches, Alaska (Clark); Kadiak (Nutting); San Diego (Torrey); Bering strait and Bering island (Jäderholm); near Shnmagin islands and east of Afognak island, 238 fathoms; Puget sound (Nutting); widely distributed in the Queen Charlotte and Vancouver island region (Fraser); point Loma, one station off Oakland and two in the northern portion of the upper section of San Francisco bay; off North head, Wash.; many further records in the Queen Charlotte and Vancouver island regions; off McArthur reef, Sumner strait, point Sherman, Lynn canal, Juneau, Berg inlet, Kadiak, Alaska. 7½ to 238 fathoms.

Campanularia volubilis (Linnaeus)
Plate 14, Fig. 63

Sertularia volubilis LINNAEUS, Syst. Nat., 1767, p. 1311.

Campanularia volubilis FRASER, West Coast Hyd., 1911, p. 34.

Vancouver Island Hyd., 1914, p. 141.

NUTTING, Am. Hyd., III, 1915, p. 31.

FRASER, Hyd. from west coast of V.I., 1935, p. 123.

Hyd. Distr. in vicinity of Q.C.I., 1936, p. 123.

Trophosome.—Stems unbranched serving as pedicels, arising from a stolon that may be plain or twisted; pedicels slender, spirally twisted or annulated; hydrothecae tubular, narrow, and deep; margin with about 10 rounded teeth which may be very shallow so that the margin appeared sinuous.

Gonosome.—Gonangia flask-shaped, with long narrow neck, borne on the stolon by means of short annulated pedicels.

Distribution.—San Pedro, Tomales bay, San Diego (Torrey); Bare island (Hartlaub); Banks island, Ucluelet, San Juan archipelago; Northumberland channel, Dodds narrows; Massett harbour (Fraser); Bering sea (Hartlaub); San Pedro and Dillon's beach, Cal. Low tide to 30 fathoms.

Genus CLYTIA

Trophosome.—Stem unbranched or irregularly branched.

Gonosome.—Gonophores producing free medusae, somewhat spherical, with four tentacles at time of liberation.

KEY TO SPECIES

A. Stem usually much branched
    a. Stem and main branches fascicled........*C. universitatis*
    b. Stem simple
        a. Gonangium smooth
        b. Gonangium corrugated
            1. Hydrotheca small, with 8 sharp teeth...*C. minuta*
            2. Hydrotheca larger, with 10-14 teeth ..*C. edwardsi*
B. Stem unbranched or but slightly branched
    a. Hydrothecal margin entire....................*C. bakeri*

b. Hydrothecal margin toothed
    a. Teeth bicuspidate.....................*C. longitheca*
    b. Teeth simple
        1. Teeth keeled
            i. 8 to 10 teeth..................*C. kincaidi*
            ii. 12 to 14 teeth...............*C. hendersoni*
        2. Teeth not keeled
            i. Hydrotheca cylindrical, 10-12 teeth........
                        ....*C. cylindrica*
            ii. Hydrotheca broadly campanulate, 12-14 teeth
                        ....*C. johnstoni*
            iii. Hydrotheca deeply campanulate
                x. Hydrotheca small, 7 teeth...........
                      ....*C. inconspicua*
                xx. Hydrotheca larger, 10-12 teeth.......
                      ....*C. hesperia*

Clytia attenuata (Calkins)
Plate 14, Fig. 64

Campanularia attenuata CALKINS, Hyd. from Puget Sound, 1899,
p. 350.
Clytia attenuata FRASER, West Coast Hyd., 1911, p. 34.
                Vancouver Island Hyd., 1914, p. 142.
      NUTTING, Am. Hyd., III, 1915, p. 60.

Trophosome.—Slender stems arising from a slender flexuous stolon; few, if any, branches; pedicels long, similar to main stem, given off at a sharp angle, passing out in the same general direction as the stem. Stem annulated above the origin of each pedicel, each pedicel annulated at the base and below the hydrotheca; hydrotheca becoming larger towards the margin; the margin with 9 or 10 teeth, not very deeply cut.

Gonosome.—"Gonotheca large, borne on short, ringed stalk on the parent stem, just above the axil of the branches; smooth, oval and with a terminal aperture. The blastostyle, as a rule, bears three medusae, the oldest of which are provided with a well marked manubrium and four tentacles; the diaphragm is a simple partition with down-turned edge at the aperture" (Calkins).

Distribution.—Port Townshend, Scow bay (Calkins); Goat island, San Francisco bay, 3 fathoms.

Clytia bakeri Torrey
Plate 14, Fig. 65

Clytia bakeri TORREY, Hyd. of San Diego, 1904, p. 16.
    FRASER, West Coast Hyd., 1911, p. 34.
    NUTTING, Am. Hyd., III, 1915, p. 59.

Trophosome.—Unbranched geniculate stems, with a maximum height of 30 mm., arising in clusters from a stoloniferous network, attached to the shell of the bivalve *Donax.* The basal portion of the stem, devoid of hydrothecae, appears to be definitely divided into rectangular internodes, shorter at the base and increasing in length towards the first hydrotheca; the hydrothecae are given off regularly alternately, producing a distinct resemblance to *Obelia geniculata.* The pedicels are short, usually with two annulations, giving here again the appearance of internodes, but not so pronounced as in the basal portion of the stem. Hydrothecae with straight walls, tapering rather rapidly from base to margin; length and breadth nearly equal; margin entire.

Gonosome.—Gonangia axillary, single or in pairs, almost sessile, obovate, with a small but usually distinct collar. Medusa buds numerous in each gonangium.

Distribution.—Pacific beach, San Diego, in the surf (Torrey); Pacific beach, Long beach and Torrey Pines beach, near La Jolla, Cal. (Nutting); near San Bruno light in the lower section of San Francisco bay and outside the Golden Gate, Balboa on Pismo clam. 6½ to 39 fathoms.

Clytia cylindrica Agassiz
Plate 14, Fig. 66

Clytia cylindrica AGASSIZ, Contr. Nat. Hist. U.S., IV, 1862, p. 306.
    FRASER, Vancouver Island Hyd., 1914, p. 142.
    NUTTING, Am. Hyd., III, 1915, p. 58.
    FRASER, Hyd. Distr. in vicinity of Q.C.I., 1936, p. 123.

Trophosome.—Stem unbranched; the slender pedicel is annulated proximally and distally; hydrotheca cylindrical, at least twice as deep as wide, suddenly constricted at the base where the diaphragm appears inside, the part below the diaphragm being little larger than the end of the pedicel. Teeth 10-12, sharp-pointed and rather deeply cut.

Gonosome.—Gonophores given off from the stolon, or occasionally from the pedicel, supported on short pedicels with one or two annulations; gonangium smooth, oblong, or slightly obovate, narrowing slightly just below the brim.

Distribution.—Off point Richardson, near Friday Harbor; western Skidegate narrows, Houston Stewart channel, off Rose harbour (Fraser). Low tide to 30 fathoms.

Clytia edwardsi (Nutting)
Plate 14, Fig. 67

Campanularia edwardsi NUTTING, Hyd. of Woods Hole, 1901, p.346.
Clytia edwardsi FRASER, West Coast Hyd., 1911, p. 34.
Vancouver Island Hyd., 1914, p. 143.
NUTTING, Am. Hyd., III, 1915, p. 60.
FRASER, Hyd. from west coast of V.I., 1935, p. 144.
Hyd. Distr. in vicinity of Q.C.I., 1936, p. 123.

Trophosome.—Stem unbranched or with few or many irregularly placed branches; the stolon often forms a complicated network on *Fucus*, worm tubes, *etc.*, but at other times passes along in a regular direction. When the whole stem consists of a single pedicel, it is long and slender, annulated at the base and below the hydrotheca; when there is only one branch, it turns abruptly in the direction of the stem, almost at its base; it is also long and slender, overreaching the main stem, the branching often being as long as the main stem or longer. On it the annulations are similarly arranged to those on the main stem; there are no annulations on the main stem immediately above where the branch is given off. When other branches are given off, they bear a similar relation to the branch from which they spring as the first branch does to the stem; thus producing a loose cymose appearance. The whole colony may reach a height of 25 or 30 mm. The hydrothecae are usually large but vary extensively in size; deeply campanulate, with 10-14 deeply cut teeth, slender but rounded at the tip.

Gonosome.—Gonangia grow from the stolon, the axils of the branches, or directly from the stem or branches; oblong or oval, corrugated, varying much in size and in the number of the corrugations; they are borne on short pedicels with 2 or 3 annulations.

Distribution.—San Diego (Torrey); Port Townshend (Calkins); widely and extensively distributed in the Vancouver island and Puget sound region; off Clayuquot sound; numerous locations in the Queen Charlotte island region (Fraser); San Diego, Avalon, San Clemente, several locations in the middle section of San Francisco bay, in an area reaching from the Golden Gate to Oakland and Richmond; Hoodsport and Alki point, Wash.; east of Pillar bay, Q.C.I., off cape James, Hope island. Low tide to 50 fathoms.

Clytia hendersoni Torrey
Plate 14, Fig. 68

Clytia hendersoni TORREY, Hyd. of San Diego, 1904, p. 18.
FRASER, West Coast Hyd., 1911, p. 35.
NUTTING, Am. Hyd., III, 1915, p. 62.

Trophosome.—Colonies up to 5 cm. high; stems unbranched or with branches similar to the main stem, geniculate, annulated above each pedicel; pedicels short, annulated throughout, arising from a definite knee and running closely parallel to the main stem. Hydrotheca longer than broad, with 12-14 sharply pointed, keeled teeth on the margin; a line runs towards the base from each indentation.

Gonosome.—"Gonangia with wide mouths, widest in distal half, tapering, 3 times as long as broad, with wavy contours but not annulated. Pedicels short, with three or four annulations. Usually 3 or 4 medusae in each gonangium, each with four tentacles and without gonads" (Torrey).

Distribution.—San Diego, 3 fathoms (Torrey); San Pablo, off Goat island in San Francisco bay, 5-8 fathoms.

Clytia hesperia (Torrey)
Plate 14, Fig. 69

Campanularia hesperia TORREY, Hyd. of San Diego, 1904, p. 12.
FRASER, West Coast Hyd., 1911, p. 31.
NUTTING, Am. Hyd., III, 1915, p. 36.

Trophosome.—Stems unbranched, serving as pedicels, arising from an undulating but not distinctly annulated stolon; short, erect, rigid, either entirely annulated, or with one or more intervals not annulated. Hydrothecae deeply campanulate, increasing in diameter from base to margin; margin with 11 or 12 sharply-pointed, distinctly cut teeth. Hydranth with 22-24 tentacles.

Gonosome.—(Not previously described.) Gonangia large, obovate, almost sessile on the stolon, wavy, not corrugated, with a very distinct collar.

Distribution.—La Jolla, Cal., between tides (Torrey); San Pedro, Cal.

Clytia inconspicua (Forbes)
Plate 15, Fig. 70

Thaumantias inconspicua FORBES, Br. Naked-eyed Medusae, 1848,
p. 52.
WRIGHT, Quart. Jour. Micr. Sc., 1862,
p. 221.
Campanularia inconspicua CALKINS, Puget Sd. Hyd., 1899, p. 349.
Thaumantias inconspicua FRASER, West Coast Hyd., 1911, p. 40.
Clytia inconspicua FRASER, Vancouver Island Hyd., 1914, p. 144.

Trophosome.—Colony small, stem usually unbranched; pedicel short and slender, annulated or wrinkled throughout, or with a small portion towards the centre smooth; hydrotheca small with 7 blunt but distinctly cut teeth.

Gonosome.—Gonangia borne on the stolon by means of short, annulated pedicels, obovate, smooth; aperture terminal, large.

Distribution.—Puget sound (Calkins); San Juan archipelago; Banks island, Departure bay, Whaleboat passage (Fraser).

Clytia johnstoni (Alder)
Plate 15, Fig. 71

Campanularia johnstoni ALDER, Ann. and Mag. N.H., (2), XVIII,
1856, p. 359.
Clytia johnstoni FRASER, West Coast Hyd., 1911, p. 36.
Vancouver Island Hyd., 1914, p. 146.
NUTTING, Am. Hyd., III, 1915, p. 54.
FRASER, Hyd. of Eastern Canada, 1918, p. 345.
Hyd. Distr. in vicinity of Q.C.I., 1936,
p. 124.

Trophosome.—Stems unbranched or sometimes with a single branch; pedicels annulated proximally and distally; hydrotheca broadly campanulate, depth and width nearly equal; margin with 12-14 triangular teeth that may be sharp or slightly rounded.

Gonosome.—Gonophores growing either from the stem or the stolon, attached by short annulated pedicels; rather small as compared with others in the same genus; oval or oblong, truncate, corrugated; opening not large.

Distribution.—Port Etches, Alaska (Clark); Puget sound (Calkins); Oakland creek (Torrey); north of Gabriola island, Gabriola pass; off Rose harbour in Houston channel (Fraser); Queen Charlotte sound, off Sadler point, Graham island; off McArthur reef and off Shingle point in Sumner strait, Alaska.   15 to 240 fathoms.

Clytia kincaidi (Nutting)
Plate 15, Fig. 72

Campanularia kincaidi NUTTING, Alaska and Puget Sd. Hyd., 1899,
p. 743.
FRASER, West Coast Hyd., 1911, p. 31.
Clytia kincaidi FRASER, Vancouver Island Hyd., 1914, p. 146.
Campanularia kincaidi NUTTING, Am. Hyd., III, 1915, p. 39.
Clytia kincaidi FRASER, Hyd. from west coast of V.I., 1935, p. 144.
Hyd. Distr. in vicinity of Q.C.I., 1936,
p. 124.

Trophosome.—Stem unbranched, pedicels slender, annulated proximally and distally and sometimes with 2 or 3 annulations medially placed; hydrotheca small, tubular, long for the width, ribbed lengthwise, the line passing down for some distance from each tooth; teeth 8 to 10, distinctly and sharply pointed.

Gonosome.—Gonangium oval or obovate, gradually increasing in size from the base for about three-fourths of its length and then narrowing slightly; the pedicel is much longer than usual in *Clytia*, having as many as 7 annulations.

Distribution.—Puget sound (Nutting); Dodds narrows; Nanoose bay, off Clarke rock, Gabriola pass, off Matia island, Friday Harbor; off Bajo reef; Rennell sound, off Rose harbour in Houston Stewart channel (Fraser); off Bull harbour, Nawhitti bar and cape James, all near Hope island, Queen Charlotte sound; off McArthur reef and Shingle island in Sumner strait, Alaska.   8 to 240 fathoms.

Clytia longitheca (Fraser)
Plate 15, Fig. 73

Campanularia longitheca FRASER, Vancouver Island Hyd., 1914,
p. 137.
Clytia longitheca FRASER, Ditto, Footnote.

Trophosome.—Stems unbranched, serving as the pedicels for the hydranths; pedicels long and slender, with several annulations at the base of the hydrotheca, fewer at the base of the pedicels and in some cases, 3-5 towards the centre; hydrotheca very long, tapering very gradually from margin to base; the 9-10 teeth are deeply cut and each is provided with two distinct cusps.

Gonosome.—Gonangium attached to the stolon by a short pedicel with three annulations; long and slender, 1.25 mm. long and 0.3 mm. in greatest width; the base is narrow and from this the gonangium gradually increases in size for the proximal third of the length, after which it is practically uniform; distaI end sharply truncate, with the opening occupying less than one-third of the surface; walls smooth; five medusae in each gonangium.

Distribution.—Nanoose bay, Departure bay, east of Protection island, near Round island Whaleboat passage (Fraser); Mare island, Goat island, and near Oakland, in San Francisco bay, 10-12 fathoms.

Clytia minuta (Nutting)
Plate 15, Fig. 74

Campanularia minuta NUTTING, Hyd. of Woods Hole Region, 1901, p. 345.

Clytia minuta FRASER, New England Hyd., 1912, p. 44.
NUTTING, Am. Hyd., III, 1915, p. 61.
FRASER, Hyd. from Q.C.I., 1935, p. 507.

Trophosome.—Stems and pedicels usually long and very slender and as the pedicels and branches leave the stem, they turn abruptly upward, side by side with the main stem; consequently, although the colony may reach a height of 2 cm. or more, the spread is insignificant. So many colonies grow close together that at first glance one would not observe the extreme slenderness of the colony. Annulation is often carried to the extreme, so that there is scarcely any part of the stem, branch, or pedicel that is not annulated or at least wavy. In no case is there any large portion free of annulations. Hydrotheca small, rather broadly campanulate; with 8 sharp teeth on the margin.

Gonosome.—Gonangium growing from the stolon or from the stem; oval or obovate, with strong transverse corrugations.

Distribution.—In tow net, near the surface, in the western portion of Houston Stewart channel (Fraser).

Clytia universitatis Torrey
Plate 15, Fig. 75

Clytia universitatis TORREY, Hyd. of San Diego, 1904, p. 19.
FRASER, West Coast Hyd., 1911, p. 36.
NUTTING, Am. Hyd., III, 1915, p. 61.

Trophosome.—Colony unusually large for this genus, up to 20 cm. in height; stem and larger branches fascicled; many large branches, irregularly arranged; hydranth pedicels given off irregularly, long and annulated throughout.   The hydrotheca is deeply campanulate, increasing in diameter very slightly from base to margin; margin with 12-15 well marked, pointed teeth.

Gonosome.—Gonangia most commonly borne on the stem but they may also be present on the branches or pedicels; sessile or almost so, oblong-oval, with the distal end sharply truncate; walls somewhat sinuous.

Distribution.—San Diego and San Pedro, 10-18 fathoms(Torrey); south Coronado island, off Ferry wharf and Goat island, San Francisco bay.   9-13 fathoms.

Genus EUCOPELLA

Trophosome.—Stem unbranched, arising from an anastomosing stolon; hydrothecae with very thick walls and entire margin.

Gonosome.—Gonophores produce large medusoid structures, devoid of mouth or digestive cavity.

KEY TO SPECIES

A.   Hydrothecal pedicel smooth or nearly so
    a.   Gonangium large, obovate, truncate........*E. caliculata*
    b.   Gonangium laterally compressed...........*E. compressa*
B.   Hydrothecal pedicel wavy or annulated.............*E. everta*

Eucopella caliculata (Hincks)
Plate 15, Fig. 76

Campanularia caliculata HINCKS, Ann. and Mag. N.H., (2), XI, 1853, p. 178.
Eucopella caliculata FRASER, West Coast Hyd., 1911, p. 36.
                      Vancouver Island Hyd., 1914, p. 147.
Orthopyxis caliculata NUTTING, Am. Hyd., III, 1915, p. 64.
Eucopella caliculata FRASER, Hyd. of Eastern Canada, 1918, p. 347.
                      Hyd. from west coast of V.I., 1935, p. 144.
                      Hyd. Distr. in vicinity of Q.C.I., 1936, 124.

Trophosome.—Stem unbranched, serving as the pedicel of the hydranth, varying in length, slightly wavy or annulated, with a distinct double annulation below the hydrotheca; hydrotheca with very thick wall and entire margin.

Gonosome.—Gonangium large, irregularly obovate; the distal end regularly rounded or somewhat truncate, attached to the stolon by means of a short pedicel. Within the gonangium are two medusoid structures, one large, occupying the greater portion of the space, and a much smaller one below. These are elongated oval, and when liberated are devoid of mouth and digestive cavity.

Distribution.—Point Wilson, Port Townshend, Bremerton, Wash. (Calkins); Yakutat, Alaska (Nutting); Bering sea (Jäderholm); San Juan archipelago; off Massett, Dodds narrows, Gabriola pass, off Matia island, off Fossil island, Friday Harbor, off point Richardson, Coupeville, point Grenville; near Maquinna point; entrance to Big bay (Fraser); near Shag rock in the lower section of San Francisco bay; Channel islands, Puget sound; off McArthur reef in Sumner strait, Alaska. Low tide to 40 fathoms.

Eucopella compressa (Clark)
Plate 16, Fig. 77

Campanularia compressa CLARK, Alaskan Hyd., 1876, p. 214.
Clytia compressa NUTTING, Hyd. of the Harriman Exp., 1901,
p. 170.
FRASER, West Coast Hyd., 1911, p. 37.
Orthopyxis compressa NUTTING, Am. Hyd., III, 1915, p. 65.

Trophosome.—"Colony consisting of unbranched pedicels springing from a creeping rootstock. Pedicels and rootstocks not regularly annulated, but with greatly thickened perisarc. Pedicels sometimes attaining a height of 6 mm., smooth, with usually a globular annulation just below the hydrotheca. Hydrothecae sometimes without greatly thickened walls and triangular in shape. Others have excessively thickened walls, so that the outline becomes ovoid, the inner surface being typically campanulate in outline. The margin is always even, never regularly dentated. When the walls are greatly thickened, the basal chamber assumes the outline of a bell handle" (Nutting).

Gonosome.—Gonangia very much compressed laterally, being often as broad as long, when viewed from the broad side. Medusae in the gonangia, with four radial canals and four tentacles (condensed from Nutting's description).

Distribution.—Shumagin islands, Alaska, 6-20 fathoms (Clark); Orca, Alaska (Nutting); San Diego, 5 fathoms, San Pedro, 3 fathoms (Torrey).

Eucopella everta (Clark)
Plate 16, Fig. 78

Campanularia everta CLARK, Hyd. of Pacific Coast, 1876, p. 253.
Eucopella everta FRASER, West Coast Hyd., 1911, p. 37.
Campanularia everta FRASER, Vancouver Island Hyd., 1914, p. 133.
Orthopyxis everta NUTTING, Am. Hyd., III, 1915, p. 67.
Eucopella everta FRASER, Hyd. of west coast of V.I., 1935, p. 144.
                Hyd. Distr. in vicinity of Q.C.I., 1936,
                    p. 124.

Trophosome.—Stems unbranched, arising from a reticulated stolon; pedicels irregularly annulated or wavy throughout, with a distinct double annulation below the hydrotheca; hydrothecae very variable; the wall may be quite thick or comparatively thin but even at the thinnest it is thicker than that of the majority of the campanularians; the margin is sometimes strongly everted and at other times not everted in the least; the margin may be perfectly even, slightly crenulated, or distinctly wavy.

Gonosome.—Gonangia are borne on the stolon by means of short pedicels that may have one or two annulations; the male gonangia are smaller than the female but are of the same shape, broadly oval in one plane and more oblong in the other; surface smooth or with large, shallow corrugations; distal end rather truncate with the opening occupying but a small portion; in the female, each sporosac becomes extended as an acrocyst.

Distribution.—San Diego (Clark); Catalina island, 42 fathoms, San Diego, 1-24 fathoms, Pacific Grove (Torrey); Port Renfrew, Departure bay; off Nootka island, Bajo reef, off Indian village, Esperanza inlet; Danger rocks at eastern entrance to Houston Stewart channel, entrance to Flamingo harbour, western portion of Houston Stewart channel, Massett harbour (Fraser); San Diego, Santa Cruz, Avalon, Catalina island, San Pedro, near Pinos light, Pacific Grove, Monterey, Cal.; off San Pablo and Oakland in San Francisco bay; off Yaquina point and Heceta head, Ore. 7 to 68 fathoms.

Genus GONOTHYRAEA

Trophosome.—Stem branched; hydrothecae campanulate with thin walls.

Gonosome.—Reproduction by fixed medusiform sporosacs, furnished with tentacles, that at maturity become extra-capsular, remaining attached until their contents are discharged.

KEY TO SPECIES

A. Hydrotheca with deeply cut, acute teeth . . . .*G. gracilis*
B. Hydrotheca with castellated margin . . . .*G. clarki*
C. Hydrotheca with entire margin . . . .*G. inornata*

Gonothyraea clarki (Marktanner-Turneretscher)
Plate 16, Fig. 79

Gonothyraea hyalina CLARK, Alaskan Hyd., 1876, p. 215.

Laomedea (Gonothyraea) clarki MARKTANNER, Hyd. von Ostspitz-
bergen, 1895, p. 408.

Gonothyraea clarki TORREY, Hyd. of the Pacific coast, 1902, p. 55.

FRASER, West Coast Hyd., 1911, p. 37.

Vancouver Island Hyd., 1914, p. 148.

NUTTING, Am. Hyd., III, 1915, p. 71.

FRASER, Hyd. as a food supply, 1933, p. 260.

Hyd. Distr. in vicinity of Q.C.I., 1936,
p. 125.

Trophosome.—Colonies branched; stem and branches slender, internodes long; main stem annulated above each of the branches; branches annulated at the base and above the origin of small branches and pedicels; hydrothecal pedicels arranged alternately, the branches slightly geniculate where they are given off, short and usually annulated throughout; hydrotheca deeply campanulate, narrowing very slightly in the distal half and somewhat more in the proximal; margin with 10-12 sharply truncated teeth, giving a castellated appearance to the hydrotheca.

Gonosome.—Gonangium oval or oblong, usually growing in the axils of branches or pedicels but occasionally taking the place of hydrothecae; medusoids four or five in each gonangium.

Distribution.—Semidi islands to Nunivak island, Alaska(Clark); Catalina island, 42 fathoms, San Diego, 1-24 fathoms, Pacific Grove (Torrey); Bare island (Hartlaub); Departure bay, San Juan archipelago; Departure bay, Nanaimo, Dodds narrows, Gabriola reefs, Friday Harbor: in the stomach of *Somateria spectabilis* and of *Histrionicus histrionicus*, in Hooper bay, Alaska, and in *Nyroca americana* in Tomales bay, Cal.; west of Horn island, in Tasoo harbour, western and eastern narrows in Skidegate channel, Houston Stewart channel at the western end and off Rose harbour, Massett harbour (Fraser); San Diego, Santa Cruz, Avalon, Catalina, San Pedro, Pacific Grove, Monterey, off San Pablo and Oakland in San

Francisco bay; off Yaquina point and Heceta head, Ore.   Low tide to 68 fathoms.

Gonothyraea gracilis (Sars)
Plate 16, Fig. 80

Laomedea gracilis SARS, Beretning om en Zool. Reise, 1851, p. 18.
Gonothyraea gracilis FRASER, Vancouver Island Hyd., 1914, p. 148.
          NUTTING, Am. Hyd., III, 1915, p. 70.
          FRASER, Hyd. from west coast of V.I., 1935,
                    p. 124.
                    Hyd. Distr. in vicinity of Q.C.I., 1936,
                    p. 144.

Trophosome.—Colony irregularly branched; stem, branches and pedicels long and slender; branches and pedicels bend abruptly near the origin and pass upward in the same direction as the main stem; stem with several annulations at the base and above the origin of each branch and pedicel; each pedicel with several annulations at the base and below the hydrotheca; hydrotheca long for its width, cylindrical for the upper half or two-thirds and gradually tapering to the base; teeth 10-14, deeply cut and rather sharp.

Gonosome.—Gonophores borne on the stolon and on the stem, with distinctly annulated pedicels; gonangia oblong-oval, often flaring a little at the rim; each gonophore bears four or five sporosacs.

Distribution.—Departure bay, west of Hammond bay, off West rocks, off Snake island, north of Gabriola island, Northumberland channel, Dodds narrows, Gabriola pass, off Matia island, Friday Harbor; off Clayoquot sound, south of Flores island, off Bajo reef; north of Marble island, northwest and south of Gospel island in Rennell sound (Fraser); off Klashwan point, off cape James, Hope island; Dewey anchorage, Etolin island, Alaska.   5 to 75 fathoms.

Gonothyraea inornata Nutting
Plate 16, Fig. 81

Gonothyraea inornata NUTTING, Hyd. of the Harriman Exp., 1901,
                    p. 175.
          FRASER, West Coast Hyd., 1911, p. 37.
                    Vancouver Island Hyd., 1914, p. 149.
          NUTTING, Am. Hyd., III, 1915, p. 72.
          FRASER, Hyd. Distr. in vicinity of Q.C.I.,
                    1936, p. 124.

Trophosome.—Colony reaching a height of 50 mm.; main stem dividing near the base into several branches that pass directly upward, usually without any further branching; hydranth pedicels given off alternately in the same plane; branches annulated, usually three annulations, above the origin of the pedicels.   There is a slight tendency to geniculation.   Pedicels short, annulated throughout; hydrotheca funnel-shaped; margin entire.

Gonosome.—Gonophores borne in the axils of the pedicels, or taking the place of pedicels, on short annulated pedicels, obconic, truncate distally, smooth or with a slight tendency to corrugation; there is but one sporosac in each gonangium.

Distribution.—Yakutat, Alaska (Nutting); Friday Harbor, north shore of Hibben island, entrance to Flamingo harbour, entrance to Big bay, in Houston Stewart channel at the western end and off Rose harbour (Fraser).   Low tide to 30 fathoms.

Genus **OBELIA**

Trophosome.—Stem branched, simple or fascicled; hydrothecae with thin walls.

Gonosome.—Reproduction by means of free medusae, that when liberated, possess more than eight marginal tentacles but no oral tentacles.   Eight interradial lithocysts are present.

KEY TO SPECIES

A.   Stem fascicled
   a.   Stem erect
      a.   Hydrotheca with entire margin............*O. plicata*
      b.   Margin with teeth having two cusps...*O. multidentata*
   b.   Stem clinging; hydrotheca with sinuous margin.*O. fragilis*
B.   Stem simple
   a.   Colony large
      a.   Hydrothecal margin entire
         1.   Main stem geniculate, with main branches coming off alternately............*O. commissuralis*
         2.   Main stem not geniculate, branches not regularly arranged......................*O. borealis*
      b.   Hydrothecal margin sinuous..........*O. longissima*
   b.   Colony small
      a.   Very much branched...................*O. griffini*
      b.   Not very much branched

1. Hydrothecal margin entire
   i. Stems unbranched or very slightly branched
       x. Pedicels borne on shoulder processes of the internodes........*O. geniculata*
       xx. No prominent internodal shoulder processes.................*O. gracilis*
   ii. Stem with alternate branches ...*O. surcularis*
2. Hydrothecal margin polyhedral .....*O. dichotoma*
3. Margin toothed
   i. Marginal teeth bimucronate
       x. Hydrotheca with distinct vertical lines..
           O. bicuspidata
       xx. Hydrotheca without vertical lines......
           ....*O. corona*
   ii. Marginal teeth broad, shallow, rounded.....
       ....*O. dubia*

Obelia bicuspidata Clarke
Plate 16, Fig. 82

Obelia bicuspidata CLARKE, Trans. Conn. Acad. Sc. III, 1876, p. 58.
   FRASER, Hyd. of Beaufort, 1912, p. 361.
   NUTTING, Am. Hyd., III, 1915, p. 80.

Trophosome.—Colony small, not much branched; main stem geniculate, annulated at the base and above each branch and pedicel; hydrothecae on short pedicels, except the terminal one, annulated throughout, standing well out from the stem; long and slender, tubular but tapering slightly to the base; margin toothed, each tooth provided with two sharp points; lines are usually present; running from the base of the indentations, lengthwise of the hydrotheca.

Gonosome.—Gonangia very small, borne in the axils of the hydrothecal pedicels, supported on short, annulated pedicels; gonangia ovate or oval, with the top truncated or, in some cases, slightly inverted at the centre; some of them are shorter than the hydrothecae.

Distribution.—Generally distributed throughout San Francisco bay. 7 to 12 fathoms.

Obelia borealis Nutting
Plate 17, Fig. 83

Obelia borealis NUTTING, Hyd. of the Harriman Expedition, 1901,
p. 174.
FRASER, West Coast Hyd., 1911, p. 38.
Vancouver Island Hyd., 1914, p. 150.
NUTTING, Am. Hyd., III, 1915, p. 85.

Trophosome.—Colony very large (Nutting reports it up to 18 inches); stem long and slender; main branches long and spreading, given off singly or in pairs; the stem often sinuous at the nodes; stem and branches with 3 or 4 annulations above the nodes; pedicels either short and entirely annulated or long and annulated only at each end. Hydrotheca funnel-shaped; margin entire.

Gonosome.—Gonangia borne in the axils on annulated pedicels; obovate, a collar present; aperture large; surface smooth or with a slight tendency to corrugation.

Distribution.—Yakutat, Bering island, Sitka, Alaska (Nutting); Ucluelet, San Juan archipelago; off Massett, Naden harbour, Bull harbour (Fraser); Berg inlet and Yakutat, Alaska.

Obelia commissuralis McCrady
Plate 17, Fig. 84

Obelia commissuralis McCRADY, Gymno. of Charleston Har., 1858,
p. 95.
FRASER, West Coast Hyd., 1911, p. 38.
NUTTING, Am. Hyd., III, 1915, p. 83.

Trophosome.—Colony large, up to 20 cm.; main stem geniculate; branches numerous; stem and branches annulated above the origin of the branches and pedicels; hydrotheca small, deeper than wide; margin entire; pedicels usually annulated throughout.

Gonosome.—Gonangia axillary, obovate, smooth, with a distinct collar.

Distribution.—San Francisco bay, between tides (Torrey); near Berkeley and Oakland in San Francisco bay. Low tide to 5 fathoms.

Obelia corona Torrey
Plate 17, Fig. 85

Obelia corona TORREY, Hyd. of San Diego, 1904, p. 14.
FRASER, West Coast Hyd., 1911, p. 38.
NUTTING, Am. Hyd., III, 1915, p. 79.

Trophosome.—Colony small, consisting of simple, flexuous stems growing from a stolon; the hydranth may be supported by a pedicel that grows directly from the stolon. The main stem is annulated above the origin of each pedicel. Pedicels, except the terminal one, short, with two or three strong annulations; the terminal pedicel much longer, with median portion not annulated; hydrotheca deeply campanulate, much longer than broad, with 10-12 well marked mucronate teeth.

Gonosome.—Gonangia borne on the stem or on the stolon, on short, annulated pedicels; oblong-ovate, with a narrowing at the distal end to form an inconspicuous collar; surface smooth or nearly so.

Distribution.—San Diego, low tide (Torrey); San Diego, Oakland creek, San Bruno light in the lower section of San Francisco bay.

Obelia dichotoma (Linnaeus)
Plate 17, Fig. 86

Sertularia dichotoma LINNAEUS, Syst. Nat., 1758, p. 812.
Obelia dichotoma FRASER, West Coast Hyd., 1911, p. 38.
$\qquad$ Vancouver Island Hyd., 1914, p. 151.
$\qquad$ NUTTING, Am. Hyd., III, 1915, p. 80.
$\qquad$ FRASER, Hyd. Distr. in vicinity of Q.C.I., 1936, p. 124.

Trophosome.—Stem slender, erect, seldom more than 25 mm. high; sometimes without branches but usually with branches irregularly given off and these may be as long as the main stem so that they have a dichotomous appearance; 3-4 annulations on stem and branches above the nodes; pedicels rather short, given off in regular alternation, usually annulated throughout; hydrotheca funnel-shaped with polyhedral margin.

Gonosome.—Gonangia borne in the axils on short annulated pedicels, obovate, smooth, with a distinct collar which tapers from base to margin; aperture small.

Distribution.—Bremerton (Calkins); Sitka, Berg inlet, Orca, Alaska (Nutting); San Pedro to Coronado islands, San Diego (Torrey); Alaska to San Diego (Nutting); Departure bay, San Juan archipelago; Alert bay, off Protection island, off Matia island, off Waldron island; western Skidegate narrows, on floating *Nereocystis* in Flamingo harbour, off Rose harbour in Houston Stewart channel, Massett harbour (Fraser); San Diego, Coronado beach, Tomales bay, Red rock, San Pedro, Elkhorn slough, Monterey; outside Golden Gate and in middle and lower sections of San Francisco bay;

off Kiwanda light, Eel river bar, Ore.; Puget sound; Sitka, Berg inlet, Glacier bay, Alaska.   Low tide to 50 fathoms.

Obelia dubia Nutting
Plate 17, Fig. 87

Obelia dubia NUTTING, Hyd. of the Harriman Exp., 1901, p. 174.

  FRASER, West Coast Hyd., 1911, p. 38.

    Vancouver Island Hyd., 1914, p. 151.

  NUTTING, Am. Hyd., III, 1915, p. 77.

  FRASER, Monobrachium parasitum, *etc.*, 1918, p. 133.

    Hyd. from west coast of V.I., 1935, p. 144.

    Hyd. Distr. in vicinity of Q.C.I., 1936, p. 124.

Trophosome.—Colony small, reaching a height of 25 mm., slightly and irregularly branched, extensively annulated; pedicels usually rather long and annulated throughout; hydrotheca large, with broad, shallow rounded teeth; vertical lines passing downward from the margin for some distance, from the indentations.

Gonosome.—Gonangia borne in the axils on annulated pedicels, pear-shaped, with a distinct but low collar and small aperture; surface almost smooth or provided with broad, shallow corrugations.

Distribution.—Orca, Alaska (Nutting); Departure bay, Dodds narrows, Ucluelet, San Juan archipelago, Queen Charlotte islands; Neck point, Five Finger islands, Snake island, Northumberland channel, Gabriola pass, Friday Harbor; western part of Houston Stewart channel, west of Rose spit; off Sydney inlet, off Nootka island (Fraser); off Alcatraz island, San Francisco bay, Monterey bay; off Yaquina light and off cape Shoalwater, Ore.; Fort Madison, Wash.   Low tide to 71 fathoms.

? Obelia fragilis Calkins
Plate 17, Fig. 88

Obelia fragilis CALKINS, Puget Sound Hyd., 1899, p. 355.

  FRASER, West Coast Hyd., 1911, p. 39.

    Vancouver Island Hyd., 1914, p. 152.

  NUTTING, Am. Hyd., III, 1915, p. 87.

Trophosome.—"Hydrocaulus clinging and never erect. Stem polysiphonic, long flexuous, branched at regular intervals; branches also comparatively long and flexuous, slightly ringed at the base and with four rings above each branch. Hydrothecae deeply bell-shaped; the chitinous periderm is exceedingly delicate and easily wrinkled or folded. Hydrothecae placed alternately at some dis-

tance apart. Margin sinuous. Stems short and annulated throughout, a large hydrotheca in the axil of each branch" (Calkins).

Gonosome.—Unknown.

Dimensions.—"Length of colony 30 mm.; length of branches
9 mm.; distance between branches 1.5 mm.; length of hydrotheca
.5 mm.; width of margin .4 mm.; number of tentacles 22-24" (Calkins).

Distribution.—Dredged in Port Townshend harbour on *Aglaophenia struthionides* (Calkins).

Obelia geniculata (Linnaeus)
Plate 17, Fig. 89

Sertularia geniculata LINNAEUS, Syst. Nat., 1767, p. 1312.
Obelia geniculata FRASER, West Coast Hyd., 1911, p. 39.
        NUTTING, Am. Hyd., III, 1915, p. 73.
        FRASER, Hyd. Distr. in vicinity of Q.C.I., 1936,
            p. 124.

Trophosome.—Stem simple, geniculate, 25 mm. high, bearing
alternate pedicels on shoulder processes of the internodes; hydrotheca as wide as deep; margin entire; pedicels annulated at each
end or throughout, usually curved away from the stem.

Gonosome.—Gonangium axillary, oval or slightly obovate; terminal collar present.

Distribution.—San Francisco, Catalina island, Coronado island,
low tide to 42 fathoms (Torrey); off Rose harbour in Houston
Stewart channel (Fraser); off Block island, Coronado, La Jolla, San
Francisco bay, outside the Golden Gate; Yaquina point, Ore.   Low
tide to 46 fathoms.

Obelia gracilis Calkins
Plate 18, Fig. 90

Obelia gracilis CALKINS, Puget Sound Hyd., 1899, p. 353.
        FRASER, West Coast Hyd., 1911, p. 39.
            Vancouver Island Hyd., 1914, p. 152.
        NUTTING, Am. Hyd., III, 1915, p. 87.
        FRASER, Hyd. from west coast of V.I., 1935, p. 144.

Trophosome.—Colony small, stem slender, reaching a height of
20 mm.; either entirely without branches or with few small branches;
stem somewhat zigzag, with one or two hydranth pedicels given off
at each bend; a branch and one hydranth pedicel may be given off
at each of the lower nodes, a little higher up, two hydranth pedicels,
one much longer than the other but bearing a smaller hydrotheca,

while at the more distant nodes there is usually but one pedicel to the node; side of the hydrotheca slightly convex; margin entire.

Gonosome.—Gonangia take the place of the lower hydrothecae; generally but one hydrotheca at a node is replaced but occasionally both of them are. The gonangia are rather slender, increasing in size from the base upwards; either smooth or somewhat irregular, but scarcely corrugated; the distal portion ends in a distinct collar.

Distribution.—Scow bay, Port Townshend (Calkins); San Juan archipelago, Port Townshend; near Maquinna point, Queen's cove (Fraser).

Obelia griffini Calkins
Plate 18, Fig. 91

Obelia griffini CALKINS, Puget Sound Hyd., 1899, p. 357.

           FRASER, West Coast Hyd., 1911, p. 39.

           Vancouver Island Hyd., 1914, p. 153.

Trophosome.—Stems much branched, giving the colony a decidedly bushy appearance although it seldom reaches a height of 50 mm., and is often not more than half that; stem and branches annulated above the nodes; the shorter pedicels are annulated throughout and sometimes the longer ones are also, but at other times there is a smooth place in the centre; hydrotheca campanulate with convex sides; margin entire.

Gonosome.—Gonangia are borne in the axils on annulated pedicels; more slender than usual in species of *Obelia*, very gradually increasing in diameter from base to apex; smooth or slightly wavy; collar distinct.

Distribution.—Puget sound (Calkins); Ucluelet, Departure bay, Dodds narrows, Gabriola pass, Porlier pass, off Sucia islands, Friday Harbor, Port Townshend, West Seattle (Fraser).

Obelia longissima (Pallas)
Plate 18, Fig. 92

Sertularia longissima PALLAS, Elench. Zooph., 1766, p. 119.

Obelia longissima FRASER, West Coast Hyd., 1911, p. 39.

           Vancouver Island Hyd., 1914, p. 153.

    NUTTING, Am. Hyd., III, 1915, p. 85.

    FRASER, Hyd. as a food supply, 1933, p. 260.

           Hyd. from west coast of V.I., 1935, p. 144.

           Hyd. Distr. in vicinity of Q.C.I., 1936, p. 124.

Trophosome.—Stem filiform, of great length, sometimes reaching 50 or 60 cm.; much branched, branches alternate; stem usually sinuous where the branches are given off; stem horn colour, or in old specimens, quite black, annulated at base and above each node as are also the branches; pedicels short and annulated throughout or longer and annulated at each end; hydrotheca campanulate; margin wavy but sometimes the waves are so shallow as to be almost imperceptible.

Gonosome.—Gonangia in the axils, oval, with a distinct collar and rather small aperture; usually smooth but sometimes slightly corrugated.

Distribution.—Unalaska (Clark); the commonest shallow water campanularian in the Queen Charlotte, Vancouver island and Puget sound region; in the stomachs of *Charitonetta albeola, Somateria spectabilis, S. v-nigra,* and *Melanitta deglandi,* in Alaska and British Columbia areas (Fraser); Sessions basin, San Pedro; outside the Golden Gate and several locations in each of the three sections of San Francisco bay; off Heceta head, Ore.; numerous locations in Puget sound, Vancouver island region, Queen Charlotte islands, northern British Columbia coast and Alaska. Low tide to 70 fathoms.

? Obelia multidentata Fraser
Plate 18, Fig. 93

Obelia multidentata FRASER, Vancouver Island Hyd., 1914, p. 154.

Trophosome.—Colony much branched; main stem and larger branches show a tendency to fasciculation; branches given off alternately with some degree of regularity but they are not all in the same plane; pedicels on the more distal branches usually short and annulated throughout; extra pedicels often occur in the axils of the shorter ones as well as in the axils of the branches, which are much longer than the regular pedicels and annulated at each end only. On the stem and branches there are 2 or 3 annulations above each node; hydrotheca large, with numerous teeth, 20-24, that are very distinctly cut and either cut off squarely at the tip so that there are sharp points where the ends meet the sides, or they may be more rounded, so that the sharp points are not so noticeable; striae pass downward from the margin at the bases of the indentations between the teeth. The fluting is so distinct that it appears as though the hydrotheca were made of separate segments, with adjacent margins turned inward together.

Gonosome.—Unknown.

Distribution.—Friday Harbor (Fraser).

Obelia plicata Hincks
Plate 18, Fig. 94

Obelia plicata HINCKS, Br. Hyd. Zooph., 1868, p. 159.

               FRASER, West Coast Hyd., 1911, p. 39.

               Vancouver Island Hyd., 1914, p. 154.

               NUTTING, Am. Hyd., III, 1915, p. 78.

               FRASER, Hyd. Distr. in vicinity of Q.C.I., 1936, p.124.

Trophosome.—Stem fascicled; branches numerous, some of them fascicled, at least in the proximal portion; stem and branches with two or three annulations above each of the nodes; pedicels long, with several annulations at each end, but there may be a shorter pedicel in the axil of the longer, and this is annulated throughout; hydrotheca campanulate, with slightly everted, entire margin.

Gonosome.—Gonangia like those of *O. dichotoma* (Marktanner-Turneretscher).

Distribution.—Puget sound, Orca and Juneau, Alaska (Nutting); Puget sound (Calkins); Departure bay, San Juan archipelago; Port Townshend; western end of eastern Skidegate narrows, western portion of Houston Stewart channel (Fraser); West Berkeley, Oakland and off Bruno light in San Francisco bay; off McArthur reef and Shingle island in Sumner strait, Point Gardiner buoy, off Admiralty island, Alaska.   Low tide to 240 fathoms.

Obelia surcularis Calkins
Plate 18, Fig. 95

Obelia surcularis CALKINS, Puget Sound Hyd., 1899, p. 355.

               FRASER, West Coast Hyd., 1911, p. 40.

               Vancouver Island Hyd., 1914, p. 155.

               NUTTING, Am. Hyd., III, 1915, p. 84.

               FRASER, Hyd. from west coast of V.I., 1935, p. 144.

Trophosome.—Colonies grow from a branched stolon; each main stem gives rise to several branches arranged in quite regular alternation; stems and large branches usually end in long tendrils that may or may not have hydrothecae at the distal ends; on the branches the internodes are often quite short; stem and branches annulated above the nodes; pedicels annulated throughout or at the ends only;

hydrothecae regularly campanulate with the margin sometimes slightly everted; margin entire.

Gonosome.—Gonangia numerous in the axils of the branches and pedicels towards the base of the colony; they are rather long, expanding gradually from base to extremity; collar low.

Distribution.—On water grasses abundant in Scow bay, Port Townshend (Calkins); Kanaka bay; Catala island on the west coast of Vancouver island (Fraser).

Family Companulinidae

Trophosome.—Colonies branched or unbranched; hydrothecae pedicellate or sessile; always operculate, the operculum formed of converging segments; hydranths with conical proboscis.

Gonosome.—Gonophores producing sporosacs or free medusae.

KEY TO GENERA

A. Hydrotheca pedicellate
   a. Hydrothecal margin distinct
      a. Hydrotheca tubular......................*Calycella*
      b. Hydrotheca turbinate
         1. Nematophores on pedicels or stolon...*Egmundella*
         2. No nematophores present..............*Lovenella*
      c. Operculum shaped like an "A" tent.......*Stegopoma*
   b. Hydrothecal margin indistinct
      a. Gonophores produce free medusae......*Campanulina*
      b. Gonophores produce sporosacs.........*Opercularella*
B. Hydrotheca sessile.............................*Cuspidella*

Genus CALYCELLA

Trophosome.—A creeping stolon gives rise to tubular hydrothecae on annulated pedicels; hydrothecal margin distinct.

Gonosome.—Gonangia borne on the stolon; acrocysts produced.

Calycella syringa (Linnaeus)
Plate 19, Fig. 96

Sertularia syringa LINNAEUS, Syst. Nat., 1767, p. 1311.
Calycella syringa FRASER, West Coast Hyd., 1911, p. 42.
                  Vancouver Island Hyd., 1914, p. 156.
                  Hyd. from west coast of V.I., 1935, p. 144.
                  Hyd. Distr. in vicinity of Q.C.I., 1936,
                       p. 124.

Trophosome.—Stem smooth, not reticulated; hydrotheca tubular; margin distinct; operculum of 8-9 segments; reduplication of margin often occurs.   There is an extreme amount of variation in the size of the hydrothecae and the length of the pedicel but in all cases the pedicel is annulated throughout.

Gonosome.—Gonangia borne on the stolon, pedicel with two or three annulations, oval or obovate; sporosacs are extruded into an acrocyst.

Distribution.—Berg inlet, Kadiak, Alaska (Nutting); Port Townshend (Calkins); Puget sound (Nutting); Bare island (Hartlaub); San Diego (Torrey); almost everywhere where dredging has been done along the coast of British Columbia and in the Puget sound region (Fraser); Dillon's beach, San Pedro, outside the Golden Gate and in several locations in all three sections of San Francisco bay; off Heceta head, Ore.; Alaska may be included with British Columbia and Puget sound.   Low tide to 150 fathoms.

Genus CAMPANULINA

Trophosome.—Stem usually branched, but not always so; hydrotheca oval or ovate; margin not distinct; segments of the operculum long and slender.

Gonosome.—Gonophores producing free medusae.

KEY TO SPECIES

A.   Colony minute; base of hydrotheca forming right angle with
    the sides.................................*C. forskalea*

B.   Colony larger; base of the hydrotheca little larger than the end
    of the pedicel............................*C. rugosa*

Campanulina forskalea (Peron et Lesueur)
Plate 19, Fig. 97

Aequorea forskalea PERON et LESUEUR, Ann. Mus. Nat., Hist.,1809,
p. 336.

Campanulina forskalea FRASER, West Coast Hyd., 1911, p. 43.
Vancouver   Island   Hyd.,   1914,
p. 157.
A new Hydractinia, *etc.,* 1922,
p. 98.

Trophosome.—Stem unbranched or slightly branched; hydrotheca oval or oblong, contracting abruptly at the base so that the base forms almost a right angle with the sides; terminating above in about 12 converging segments; hydranth with 12 tentacles.

Gonosome.—Gonangium with one medusa-bud is attached to the pedicel of the nutritive zooid by a short pedicel with one annulation; the pedicel increases in diameter distally until it passes into the gonangium without any definite constriction. The gonangium is more than twice as broad as the hydrotheca; the medusa-bud is nearly equal in length and breadth; four radial canals and four tentacle bulbs, each giving rise to a tentacle, are present while the bud is still in the gonangium.

Distribution.—San Juan archipelago: near Round island in Dodds narrows; in Taylor bay, Gabriola island, Snake island (Fraser); off Mile rock and Alcatraz island in San Francisco bay. Low tide to 9 fathoms.

Campanulina rugosa Nutting
Plate 19, Fig. 98

Campanulina rugosa NUTTING, Hyd. of the Harriman Exp., 1901, p. 176.
FRASER, West Coast Hyd., 1911, p. 44.
Vancouver Island Hyd., 1914, p. 157.

Trophosome.—Colony small, seldom over 10 mm.; stem irregularly branched or even unbranched; it gives rise to hydranth pedicels in regular alternation; commonly the pedicel appears as though it was the continuation of the stem below its origin, while the continued portion of the stem appears like a branch given off from it, this makes a distinct geniculation; where there are branches given off, the regularity is interfered with; stem, branches, and pedicels are all annulated throughout; the pedicels are short with 3 or 4 annulations; hydrotheca rather stout for the length, almost oval in shape; opercular segments 10-12, about one-third of the total length of the hydrotheca.

Gonosome.—Gonangia in the axils of the lower branches and pedicels; nearly oblong but tapering slightly towards the base and somewhat flattened at the distal end; each gonangium produces a single medusa.

Distribution.—Juneau, Alaska (Nutting); West Seattle (Fraser); Oakland (?).

Genus CUSPIDELLA

Trophosome.—Hydrothecae—sessile on a creeping stolon; tubular.

Gonosome.—Unknown.

KEY TO SPECIES

A. Hydrotheca large...............................*C. grandis*
B. Hydrotheca small...............................*C. humilis*

Cuspidella grandis Hincks
Plate 19, Fig. 99

Cuspidella grandis HINCKS, Br. Hyd. Zooph., 1868, p. 210.
        FRASER, Vancouver Island Hyd., 1914, p. 158.
                Hyd. Distr. in vicinity of Q.C.I., 1936,
                p. 124.

Trophosome.—Sessile, tubular hydrothecae grow from a regularly creeping stolon; length may reach 0.8 mm. and diameter 0.15 mm.; operculum of 8-10 segments.

Gonosome.—Unknown.

Distribution.—Rose spit, Departure bay; Rennell sound, south of Gospel island (Fraser).   20 to 70 fathoms.

Cuspidella humilis (Alder)
Plate 19, Fig. 100

Campanularia humilis ALDER, Trans. Tynes. F.C., 1862, p. 239.
Cuspidella humilis FRASER, West Coast Hyd., 1911, p. 44.
                Vancouver Island Hyd., 1914, p. 159.
                Hyd. Distr. in vicinity of Q.C.I., 1936,
                p. 124.

Trophosome.—Stolon slender; hydrotheca stout in comparison with its length, but quite minute; cylindrical, sessile; operculum of 10-12 segments.

Gonosome.—Unknown.

Distribution.—Departure bay, San Juan archipelago; off Clarke rock, north of Gabriola island, Northumberland channel, Dodds narrows, Pylades channel, Gabriola pass, Ruxton passage, Whaleboat passage, Friday Harbor; west of Rose spit (Fraser); off Goat island in San Francisco bay.   9 to 30 fathoms.

Genus EGMUNDELLA

Trophosome.—Colony branched or unbranched; conspicuous nematophores on the stolon or on the pedicels; hydrotheca turbinate; operculum sharply defined by a sinuous margin on the tube of the hydrotheca.

Gonosome.—Unknown.

Egmundella gracilis Stechow
Plate 19, Fig. 101

Egmundella gracilis STECHOW, Hyd. Deutschen Tiefsee-Exp., 1921, p. 226.

Hyd. des Mittelmeeres, *etc.*, 1923, p. 124.

FRASER, Hyd. from west coast of V.I., 1935, p. 144.

Hyd. Distr. in vicinity of Q.C.I., 1936, p. 124.

Trophosome.—Stems arising from an irregularly branched stolon; unbranched or with one or two branches; pedicels, annulated at the base, pass almost imperceptibly into the turbinate hydrothecae; operculum of 12 segments. On the pedicels or on the stolon or on both, there are spherical nematophores, contracted at the base to form short pedicels.

Gonosome.—Unknown.

Distribution.—Near Southampton light, San Francisco bay; off Clayoquot sound; Houston Stewart channel at the western end and off Rose harbour. 7 to 38 fathoms.

This species seems to closely resemble *Lovenella producta* (Sars) except for the presence of the nematophores. In the Vancouver island region some colonies have been found with, and some without, the nematophores, so much alike that otherwise they are difficult to distinguish, although those without nematophores are, in general, much larger than those with them. Since that is the case, it can scarcely be that the nematophores appear only at a certain stage of the development. It would appear that both species are present in this region. The specimens from San Francisco bay were definitely provided with nematophores and hence must be placed with *Egmundella gracilis.*

Genus LOVENELLA

Trophosome.—Colony branched or unbranched; hydrotheca turbinate; operculum sharply defined by a sinuous margin on the tube of the hydrotheca; no nematophores present.

Gonosome.—"Gonangia borne on the stems, producing free, bell-shaped medusae with eight tentacles in two sets, and four lithocysts" (Nutting).

Lovenella producta (Sars)
Plate 19, Fig. 102

Calycella producta Sars, Norges Hydroider, 1873, p. 30.
Lovenella producta Fraser, West Coast Hyd., 1911, p. 44.
Vancouver Island Hyd., 1914, p. 159.

Trophosome.—Stems radiate in all directions from an irregularly branched stolon, often densely aggregated, usually unbranched but occasionally with one or two branches which grow out almost at right angles and then turn upward to overtop the stem; pedicels, up to 6 mm., annulated at the base and more or less throughout; they pass almost imperceptibly into the turbinate hydrothecae; margin of hydrotheca very distinctly scalloped for the base of the segments of the operculum, 12 or more in number; the portion of the hydrotheca nearest to the operculum may be ribbed longitudinally; the operculum may be inverted.

Gonosome.—Unknown.

Distribution.—Dodds narrows, San Juan archipelago: Lasqueti island, Nanoose bay, Clarke rock, Departure bay, north of Gabriola island, east of Protection island, Northumberland channel, Dodds narrows, Gabriola pass, Whaleboat passage, off Matia island, Friday Harbor (Fraser).

Evidently the specimens from West Rocks and from Griffin bay and possibly some of the others reported in the Vancouver island paper, really belong to the species *Egmundella gracilis*. See note on that species.

Genus OPERCULARELLA

Trophosome.—Hydrotheca elongated-oval with no distinct margin; operculum segments long and narrow.

Gonosome.—Reproduction by sporosacs that are extruded into an acrocyst.

Opercularella lacerata (Johnston)
Plate 19, Fig. 103

Campanularia lacerata JOHNSTON, Br. Zoophytes, 1847, p. 111.
Opercularella lacerata HINCKS, Br. Hyd. Zooph., 1868, p. 160.
FRASER, Vancouver Island Hyd., 1914, p. 160.

Trophosome.—Stem short, reaching a height of 15-25 mm.; branched, some of the branches being almost as long as the main stem; stem and branches flexuous, annulated throughout; pedicels short, annulated; hydrotheca with proximal portion oval and distal portion conical, no distinct margin; segments long and slender; hydranths long and slender with 14-15 tentacles.

Gonosome.—Female gonangia oval, sessile, or on short annulated pedicels, in the axils or taking the place of hydrothecae; sporosacs extruded into an acrocyst; male gonangia similar in position to the female but more nearly cylindrical in shape.

Distribution.—On *Gonothyraea clarki*, Jesse island in Departure bay. Low tide.

Genus STEGOPOMA

Trophosome.—Hydrotheca with an operculum formed of two membranes folded lengthwise, and which come together, roof-like, with their long edges; each of them is separated from the remainder of the hydrotheca by a curved line; at each side the hydrothecal wall forms a triangular, gable-like structure, between the two opercular membranes.

Gonosome.—Levinsen, who established this genus, gave no characters for the gonosome, and no one seems to have done so since. In some cases at least, probably in all, reproduction takes place by fixed sporosacs.

Stegopoma plicatile (Sars)
Plate 19, Fig. 104

Lafoea plicatile SARS, Selsk. Forhandl., 1862, p. 31.
Stegopoma plicatile LEVINSEN, Meduser, Ctenophorer, *etc.*, 1893,
p. 36.
FRASER, West Coast Hyd., 1911, p. 45.
Vancouver Island Hyd., 1914, p. 161.

Trophosome.—Stem large, strongly fascicled, the number of tubes in the fascicle being from 3 up; only the extremities of the branchlets are simple; the hydrothecae are long, almost tubular, sometimes straight but more commonly curved; often with a short pedicel but sometimes sessile or even with lateral contact with the branch; operculum consisting of the regular two membranes meeting along a ridge, with the walls of the hydrotheca produced to form a gable and to support the operculum.

Gonosome.—Sars says the gonangia are large, cylindrical, with opening at the distal end. They are found scattered over the colony; the wall is covered with a chitinous network. Levinsen says that they are formed like a long sack, growing to the branches for the greater part of their length. Broch says that in the creeping colonies, the gonangia can be distinguished from the hydrothecae only by their shorter pedicels, while in erect colonies, the gonangia are greatly elongated, oviform, fixed to the stem by the tapering extremity. Judging from Levinsen's small scale drawing, the oval gonangium is about three times the length of the hydrotheca.

Distribution.—Bering sea (Jäderholm); off Lasqueti island, West rocks, north of Gabriola island, Gabriola reefs (Fraser).

Family **Halecidae**

Trophosome.—Hydrothecae reduced to saucer-shaped hydrophores, which usually pass without constriction into the broad, tubular pedicels; they are too small to lodge the contracted hydranths; margin entire, often flaring; reduplication common; hydrophores with a circle of bright dots just below the rim; hydranth with conical proboscis.

Gonosome.—Gonophores producing fixed sporosacs or medusoid structures; there is often a decided difference between the male and female gonangia.

KEY TO GENERA

A. No nematophores or tentacular organs present
    a. Gonophores producing typical sporosacs........*Halecium*
    b. Gonophores producing medusoid structures..*Campalecium*
B. Tentacular organs present....................*Ophiodissa*

Genus **CAMPALECIUM**

Trophosome.—As in the family; no nematophores or tentacular organs.

Gonosome.—Gonangia bearing medusoid gonophores.

Campalecium medusiferum Torrey
Plate 20, Fig. 105

Campalecium medusiferum TORREY, Hyd. of the Pacific coast, 1902, p. 48.

FRASER, West Coast Hyd., 1911, p. 46.

Trophosome.—"Stem short, 5-10 mm., sparingly and irregularly branched, rooted by a creeping stolon. Hydrotheca with strongly everted rim. Hydranth large, with low conical proboscis and 24 to 28 tentacles in one whorl" (Torrey).

Gonosome.—"Gonotheca on short peduncle arising just below a hydrotheca; about three times as long as broad; broader at distal end, which is truncate, tapering gradually to the peduncle. Orifice not determined. Gonophores 2 to 5 in linear series; medusoid, with four tentacles developing in pairs which differ in size, and a conical manubrium" (Torrey).

Distribution.—Long beach, Cal., 6 fathoms (Torrey).

Genus HALECIUM

Trophosome.—As in the family; no tentacular organs present.

Gonosome.—Gonangia usually different in the two sexes.

KEY TO SPECIES
A. Stem simple
    a. Strongly annulated
        a. Stem erect
            1. Branches few, hydrophores at the end of the branchlets...................*H. annulatum*
            2. Ultimate branchlets formed by a succession of hydrophores..................*H. speciosum*
        b. Stem creeping......................*H. corrugatum*
    b. Not strongly annulated
        a. Stem very slender, often wavy..........*H. tenellum*
        b. Stem not so slender
            1. Hydrophore pedicels long............*H. ornatum*
            2. Hydrophore pedicels short or almost absent......
              ....*H. kofoidi*
        c. Colony low, without continuous stem ...*H. pygmaeum*
B. Stem fascicled
    a. Hydrophores sessile
        a. Stem with few branches, either primary or secondary
        ....*H. articulosum*

    b. Stem much branched, making a dense colony . . . .
. . . . *H. scutum*

  b. Hydrophores pedicellate

    a. Delicate branches grow directly from strongly fascicled stem

      1. Hydrophore pedicel arising from distal end of internode . . . . *H. flexile*

      2. Hydrophore pedicel at proximal end of internode.
. . . . *H. reversum*

    b. Branches fascicled to some extent

      1. Hydrophores flaring

        i. Hydrophores strongly flaring

          I. Hydrophores extensively reduplicated . . . .
H. telescopicum

          II. Hydrophores with little or no reduplication

            x. Primary branches not much branched . . . . *H. washingtoni*

           xx. Primary branches with numerous branches

              a. Branching flabellate

                o. Gonangium circular . . . . . .
. . . . *H. wilsoni*

                oo. Gonangium oblong-oval . . .
. . . . *H. labrosum*

              aa. Branching not flabellate

                o. B r a n c h l e t s numerous throughout the whole length of the branch . . .
. . . . *H. densum*

                oo. Branchlets at distal end of branches only . . . .
. . . . *H. parvulum*

        ii. Hydrophores slightly flaring

          I. Nodes transverse, gonangium smooth . . . .
. . . . *H. beani*

          II. Nodes oblique, gonangium spiny . . . .
. . . . *H. muricatum*

      2. Hydrophores not flaring

        i. Ultimate branches in dense tufts . . *H. robustum*

        ii. Ultimate branches pinnately arranged . . . .
. . . . *H. halecinum*

Halecium annulatum Torrey
Plate 20, Fig. 106

Halecium annulatum TORREY, Hyd. of the Pacific coast, 1902, p. 49.

FRASER, West Coast Hyd., 1911, p. 46.

Vancouver Island Hyd., 1914, p. 163.

Hyd. from west coast of V.I., 1935, p. 144.

Hyd. Distr. in vicinity of Q.C.I., 1936, p. 124.

Trophosome.—Delicate stems arise from a stolon, the larger of which are branched irregularly; stem and branches annulated throughout, in some cases regularly and in others irregularly; hydrothecae arising from the stem, sessile or nearly so, from the branches on longer pedicels; reduplications common, with long pedicels between the hydrophores.

Gonosome.—Female gonangia much compressed, so that they appear almost circular in the one view and narrowly oval in the other; aperture terminal; a single gonophore with several irregular processes reaching out to the wall of the gonangium.

Distribution.—Coronado, low tide to 10 fathoms (Torrey); Port Renfrew, Ucluelet, Dodds narrows; off Long beach, off Sydney inlet; entrance to Flamingo harbour; in tow net in western portion of Houston Stewart channel and west of cape St. James, Massett harbour (Fraser); Coronado island, Wilson's cove, San Clemente, west of Aumentos rocks near point Pinos, two locations in the lower section of San Francisco bay; off North head, Ore.; Symonds point in Lynn canal, Kadiak, Alaska.   Low tide to 81 fathoms.

Halecium articulosum Clark
Plate 20, Fig. 107

Halecium articulosum CLARK, New England Hyd., 1876, p. 63.

FRASER, Vancouver Island Hyd., 1914, p. 164.

Hyd. Distr. in vicinity of Q.C.I., 1936, p. 124.

Trophosome.—Stem coarse, fascicled, primary branches scarce but long, hence the colony has a very loose appearance; ultimate branches pinnately arranged, white, as distinct from the larger branches and main stem which are dark brown; internodes short, getting shorter towards the end of the branches, where they may be as broad as long; hydrothecae sessile, alternately arranged, one distally, on each internode.

Gonosome.—Female gonangia large, obovate, borne in rows on the upper side of the branches, aperture lateral but near the distal end. "Male gonangia are oblong, subcylindrical, sessile" (Clark).

Distribution.—Jumbo channel, Wash. (Fraser); off Rose harbour in Houston Stewart channel. 30 fathoms.

Halecium beani (Johnston)
Plate 20, Fig. 108

Thoa beani JOHNSTON, Br. Zooph., 1847, p. 120.
Halecium beani HINCKS, Br. Hyd. Zooph., 1868, p. 224.
            FRASER, Can. Atlantic Fauna, 1921, p. 30.
Trophosome.—Stem and main branches fascicled; nodes oblique; hydrophore margin flaring little.

Gonosome.—Gonangia borne at the base of the hydrophores; male, regular oblong-oval; female, mitten-shaped, aperture lateral; two small hydranths are present in the aperture.

Distribution.—Off Shag rock in lower section of San Francisco bay, 6 fathoms.

Halecium corrugatum Nutting
Plate 20, Fig. 109

Halecium corrugatum NUTTING, Hyd. from Alaska and Puget sound,
            1899, p. 745.
            FRASER, West Coast Hyd., 1911, p. 47.
            Vancouver Island Hyd., 1914, p. 164.
            Hyd. from west coast of V.I., 1935,
            p. 144.
            Hyd. Distr. in vicinity of Q.C.I.,
            1936, p. 124.

Trophosome.—Colony creeping, stem bearing but one hydrophore, or branching sufficiently to bear as many as 3 or 4; these arranged very regularly on the stolon; the stolon is not annulated but the pedicels are distinctly and regularly annulated throughout; margin flaring but not very strongly.

Gonosome.—Gonangia borne directly on the stolon with scarcely a sign of pedicel; the proximal half is larger; after this the gonangium narrows gradually almost to the distal end, when the diameter is again increased; aperture occupying all the distal end; surface smooth.

Distribution.—Puget sound (Nutting); Clayoquot sound, Nanoose bay, north of Gabriola island, off Matia island, off Clayoquot sound; northwest of Gospel island, in Rennell sound, entrance to Flamingo harbour, in Houston Stewart channel near the western end and off Rose harbour, Massett harbour, outside of Massett inlet (Fraser); Jamestown bay, Cal., outside Golden Gate and near Alcatraz island and Southampton light, in the middle section of San Francisco bay; off cape James, Hope island; Symonds point in Lynn canal, Sitka, Alaska. Low tide to 60 fathoms.

Halecium densum Calkins

Plate 20, Fig. 110

Halecium densum CALKINS, Hyd. from Puget Sound, 1899, p. 343.
FRASER, West Coast Hyd., 1911, p. 47.
Vancouver Island Hyd., 1914, p. 164.
Hyd. Distr. in vicinity of Q.C.I., 1936, p. 144.

Trophosome.—Stem stout, fascicled, densely branched; one, two, or three branchlets or pedicels may arise from the same node; distally the pedicels are often in groups of three; the node is not always distinctly marked but usually shows plainly just above where the branch or pedicel is given off; branches and pedicels may be wavy in outline; hydrophores with much everted rim.

Gonosome.—(Not previously described). Large gonangia clustered on the stem near the base, shaped somewhat like a cockle shell, *i.e.*, circular or nearly so in the one direction, heart-shaped in cross section, with distinct ribbing, the ribs spreading out like a fan, each ending in a prominent projection on the margin, that at the apex becomes almost spine-like. Blastostyle much branched, a branch going out to each section on each side.

Distribution.—Bremerton (Calkins); Port Renfrew, Ucluelet, San Juan archipelago; off Massett, off Rose spit, Dodds narrows, off O'Neale island, Friday Harbor, Puget sound; off Rose harbour in Houston Stewart channel (Fraser); off Frederick island, off Sadler point, 16 miles northeast of Reef island. 15 to 30 fathoms.

Halecium flexile Allman
Plate 21, Fig. 111

Halecium flexile ALLMAN, Challenger Report, 2, vol. 23, 1888, p. 11.
FRASER, Vancouver Island Hyd., 1914, p. 165.

Trophosome.—Main stem coarse, erect, strongly fascicled; the branches coming off the main stem are simple and unbranched except that a small proximal portion may be fascicled; pinnately arranged with much regularity; each branch usually passes to its extremity without forking. The branch is divided into internodes by nodes that are nearly transverse; from the distal end of each internode is given off the pedicel for a hydrophore, this pedicel being long but showing no sign of a joint; commonly the hydrophore is reduplicated one to several times, but in each case there is a long portion intervening. The margin of the hydrophore is very little everted; hydranth large, with 12-14 tentacles.

Gonosome.—Male gonangia broadly club-shaped, with distinct pedicels arising in rows from the branches just at the point where the pedicels are given off.

Distribution.—Nanoose bay, off West rocks, Departure bay, off Matia island (Fraser); Kaison bank west of Moresby island. 110 fathoms.

Halecium halecinum (Linnaeus)
Plate 21, Fig. 112

Sertularia halecina LINNAEUS, Syst. Nat., 1767, p. 1308.
Halecium halecinum FRASER, West Coast Hyd., 1911, p. 47.
Vancouver Island Hyd., 1914, p. 165.

Trophosome.—Stem fascicled, erect, rigid; primary branches fascicled, few, long, running in the same direction as the main stem; secondary branches arranged in a regular manner, and the pedicels are arranged pinnately in these; hydrophores tubular, with rim little flaring, if at all.

Gonosome.—Gonangia borne in rows on the upper side of the branchlets; female, gradually broadening distally, main portion truncated, but one side projecting in the form of a small tube, through the aperture of which pass two hydranths; male gonangia oblong, tapering slightly towards the base.

Distribution.—Puget sound, Kadiak, Alaska (Nutting); Ucluelet, Swiftsure shoal (Fraser); off Brothers light in San Francisco bay; Berg inlet, Alaska. 7 to 25 fathoms.

Halecium kofoidi Torrey
Plate 21, Fig. 113

Halecium kofoidi TORREY, Hyd. of Pacific coast, 1902, p. 49.
FRASER, West Coast Hyd., 1911, p. 47.
Vancouver Island Hyd., 1914, p. 166.

Trophosome.—Colony with a coarse stem; main branches few and irregularly placed; secondary branches short; regularly alternate, divided into approximately equal internodes by oblique nodes; each internode usually bears on a shoulder near the distal end, a sessile hydrotheca; reduplication often takes place, each new hydrotheca appearing on a short stalk, narrower at the base, arising within the previous hydrophore. In some cases the stalked hydrophores may appear directly from the internodes. There are but few indications of annulation or waviness.

Gonosome.—Male gonangia arise by a pedicel that is little more than a constriction at the base of the gonangium, from a primary or secondary internode, near its distal end; elongated oval or obovate; female gonangia have not been observed or described.

Distribution.—Point Loma, San Diego, San Pedro, Catalina island (Torrey); San Juan archipelago (Fraser); San Diego, San Pedro, Catalina island, three locations in the upper and middle sections of San Francisco bay. 5 to 42 fathoms.

Halecium labrosum Alder
Plate 21, Fig. 114

Halecium labrosum ALDER, Ann. and Mag. N.H., (3), III, 1859,
p. 354.
FRASER, Vancouver Island Hyd., 1914, p. 167.
Hyd. Distr. in vicinity of Q.C.I., 1936,
p. 124.

Trophosome.—Stem fascicled, sparsely branched; primary branches fascicled, secondary branches pinnately arranged; these may branch again; the pedicels are borne singly or in pairs, from the distal end of the internode; three or four annulations at the proximal end of the internodes of the branches and about the same number at the base of the pedicels, which are long and tubular; margin of the hydrophore very distinctly everted; when duplication takes place the tube between the hydrophores is long.

Gonosome.—Gonangia ovate, growing in rows on the upper surface of the branches; male and female more nearly alike than in
other species of the genus.

Distribution.—North of Gabriola island, Dodds narrows; in the
western portion of Houston Stewart channel and off Rose harbour
(Fraser); off Shingle island in Sumner strait, Point Gardiner buoy
off Admiralty island, Alaska.   15 to 240 fathoms.

Halecium muricatum (Ellis and Solander)
Plate 21, Fig. 115

Sertularia muricata ELLIS and SOLANDER, Nat. Hist. Zooph., 1786,
p. 59.
Halecium muricatum FRASER, West Coast Hyd., 1911, p. 47.

Trophosome.—Stem fascicled, stout, rigid, irregularly and
densely branched; primary branches fascicled; ultimate branches
and pedicels pinnately arranged; hydrophore with margin flaring.

Gonosome.—Gonangia crowded on the branches, ovate, much
greater in the one diameter than in the other; numerous prickles on
the surface arranged in raised rows.

Distribution.—Unalaska (Clark); Orca, Alaska (Nutting); point
Gardiner buoy off Admiralty island, Symonds point in Lynn canal,
Alaska.   9 to 29 fathoms.

Halecium ornatum Nutting
Plate 22, Fig. 116

Halecium ornatum NUTTING, Hyd. of the Harriman Exp., 1901,
p. 181.
FRASER, West Coast Hyd., 1911, p. 48.

Trophosome.—"Colony parasitic, branching irregularly.   Stems
not fascicled; the stem and branches sparsely and irregularly annulated.   Pedicels long, of equal diameter throughout.   Hydrophores
with broad, everted margins, occasionally reduplicated.   Hydranths
large, with twenty-four to thirty tentacles" (Nutting).

Gonosome.—"A single apparently young gonangium was borne
on a pedicel just below the hydrophore.   It was in the form of a
truncated and deeply annulated cone.   Probably the mature gonangium would resemble that of *H. speciosum*" (Nutting).

Distribution.—Berg inlet, Glacier bay, Alaska (Nutting).

Halecium parvulum Bale
Plate 22, Fig. 117

Halecium parvulum BALE, Proc. Linn. Soc. N.S.W., 1888, p. 760.
Halecium balei FRASER, West Coast Hyd., 1911, p. 46.
Halecium parvulum BALE, Vancouver Island Hyd., 1914, p. 167.

Trophosome.—Stem and larger branches fascicled; main branches few and short; branches not numerous proximally but more so distally. In some cases near the tip of the branches there are many short branchlets or pedicels, so much so that it looks not unlike the tip of a branch of *H. densum* but the latter is stouter and more wavy. The proximal internodes are long but the distal are quite short, the nodes are oblique; occasionally the pedicel or the branchlet may be more or less annulated; margin of hydrotheca well everted.

Gonosome.—Female gonangia large, oval, compressed; orifice large, terminal; ova large; male somewhat similar in shape, but much smaller; found growing directly from the branches, sometimes in the place of hydrophores.

Distribution.—San Juan archipelago; Rose spit, Swiftsure shoal, Hammond bay, Dodds narrows, Gabriola pass, Porlier pass, off Matia island, off Sucia islands, Ruxton passage, Friday Harbor (Fraser).

Halecium pygmaeum Fraser
Plate 22, Fig. 118

Halecium pygmaeum FRASER, West Coast Hyd., 1911, p. 48.
Vancouver Island Hyd., 1914, p. 168.
Hyd. Distr. in vicinity of Q.C.I., 1936, p. 124.

Trophosome.—Colony minute, from a creeping stolon, with no continuous main stem; a single pedicel grows out from the stolon, giving rise to a hydrophore terminally; just below the hydrophore, one or two pedicels may be given off, each turning upward abruptly at the base; this may be repeated with the secondary pedicels until a series of 5 or 6 appears, but few of the colonies have so many; each pedicel has 1 to 3 annulations at the base; the hydrophore is tubular, with the margin little everted.

Gonosome.—Gonangia borne on the pedicels, similarly placed to the hydranth pedicels; the male is long, almost cylindrical but tapering slightly to the distal end and more so to the proximal; the female is obovate, with an opening on the side near the distal end, shaped like a half moon; the ova are large, 6-8 in each gonangium. Usually the gonangium is larger than the whole colony that bears it.

Distribution.—San Juan archipelago; China Hat, Friday Harbor (Fraser); off Shag rock in lower section of San Francisco bay, 6¾ fathoms, in tow net, near surface, in western portion of Houston Stewart channel.

Halecium reversum Nutting
Plate 22, Fig. 119

Halecium reversum NUTTING, Hyd. of the Harriman Exp., 1901, p. 182.
FRASER, West Coast Hyd., 1911, p. 48.
Vancouver Island Hyd., 1914, p. 168.

Trophosome.—Stem short, stout, fascicled; branches simple, arranged alternately; divided into internodes that are not annulated; each internode gives rise to a hydrophore from its proximal portion; pedicels vary much in length; hydrophore margin slightly everted.

Gonosome.—Unknown.

Distribution.—Juneau, Alaska (Nutting); north of Gabriola island (Fraser); off cape James, Hope island; Berg inlet, Glacier bay, Alaska.   50 fathoms.

Halecium robustum Nutting
Plate 22, Fig. 120

Halecium robustum NUTTING, Hyd. of Harriman Exp., 1901, p. 182.
FRASER, West Coast Hyd., 1911, p. 48.

Trophosome.—Stem stout, strongly fascicled, the numerous tubes much interwoven; there are but few large primary branches, also strongly fascicled; secondary branches much smaller but also fascicled, numerous but irregularly arranged; ultimate branches numerous, forming dense tufts, as the tubular pedicels of the hydrophores come off all sides of the branches; hydrophores not flaring, appearing as the distal portion of the pedicels; hydranths large, so much so, that in the living or well-preserved colony, the ultimate branches are almost entirely covered.

Gonosome.—Unknown.

Distribution.—Berg inlet, Glacier bay, Alaska (Nutting); Berg inlet, Glacier bay, Alaska.

Halecium scutum Clark
Plate 22, Fig. 121

Halecium scutum CLARK, Alaskan Hyd., 1876, p. 218.

FRASER, West Coast Hyd., 1911, p. 49.

Vancouver Island Hyd., 1914, p. 169.

Trophosome.—Stem stout, fascicled, much and irregularly branched; primary branches fascicled; ultimate branches much lighter in colour; divided by more or less oblique nodes into short wedge-shaped internodes, each of which has one or two hydrophores varying somewhat in position from the extreme distal end of the internode to some distance from it; hydrophores sessile; if reduplicated, the second sits quite close to the first; rather stout, very slightly everted.

Gonosome.—Gonangia ovate with an aperture projecting from the side near the distal end, although varying somewhat in position.

Distribution.—Semidi islands to Unalaska (Clark); Berg inlet and Yakutat (Nutting); San Juan archipelago, Copalis beach, Wash. (Fraser); San Pablo bay, 3 fathoms; Kadiak, Alaska.

Halecium speciosum Nutting
Plate 22, Fig. 122

Halecium speciosum NUTTING, Hyd. of the Harriman Exp., 1901, p. 181.

FRASER, West Coast Hyd., 1911, p. 49.

Trophosome.—Colony with a short, stout stem, giving rise to a very few stout branches; in some cases the secondary branches are again branched, the ultimate branches consisting of a series of hydrophores in succession, each given off from the more proximal hydrophore just below the hydrophore. These may be given off quite regularly alternating to give a geniculate effect but they are not all in the same plane and often are not of the same length; the hydrotheca is large and flaring, with a well marked row of dots.

Gonosome.—"Gonangia borne on rather long, annulated pedicels, below the hydrophores, particularly on the upper part of the colony; regularly ovoid in outline, and evenly and beautifully annulated throughout" (Nutting).

Distribution.—Yakutat, Alaska (Nutting); San Pablo bay, in the upper section, and Shag rock, in the lower section of San Francisco bay; Kadiak, Symonds point in Lynn canal, Alaska. 3 to 9 fathoms.

Halecium telescopicum Allman
Plate 23, Fig. 123

Halecium telescopicum ALLMAN, Challenger Report, Pt. II, 1888,
p. 10.
JÄDERHOLM, Hyd. Beeringsmeeres, 1907, p.4.
FRASER, West Coast Hyd., 1911, p. 49.

Trophosome.—Stem stiff, erect, fascicled, especially towards the
base; somewhat strongly branched, the branches lying in the same
plane; the distal portions of the main stem, as well as of the branches,
more slender and simple. The hydrophores appear in series in which
one hydrophore arises from within the preceding; the series is cylindrical, *i.e.*, with similar diameter throughout; there may be as many
as 12 in a series; margin of the hydrophore is thin, turned outward
and then backward. The first of the accessory segments is provided with two oblique annulations at the base.

Gonosome.—Gonangia (probably male) are elliptical, rounded at
the tip, .55-.60 mm. long, .35-.40 mm. broad; pedicel very short.

Distribution.—Bering sea (Jäderholm).

The description and the figures are taken from Jäderholm.

Halecium tenellum Hincks
Plate 23, Fig. 124

Halecium tenellum HINCKS, Ann. and Mag. N.H., (3), VIII, 1861,
p. 252.
FRASER, West Coast Hyd., 1911, p. 49.
Vancouver Island Hyd., 1914, p. 170.
Hyd. from gulf of Alaska, 1914, p. 219.
Hyd. from west coast of V.I., 1935,
p. 144.
Hyd. Distr. in vicinity of Q.C.I., 1936,
p. 124.

Trophosome.—Colony small, not over 15 mm. in height; stem
delicate, sometimes annulated or wavy; irregularly branched;
branches given off below the hydrophores, making almost a right
angle with the stem; hydrophores strongly flaring.

Gonosome.—Gonangia oval or ovate, smooth, borne at the base
of the branches or below the hydrophores.

Distribution.—San Diego (Clark); Trinity islands in gulf of
Alaska; San Juan archipelago; Rose spit, Prince Rupert, Swiftsure
shoal, Hammond bay, north of Gabriola island, Pylades channel,

Gabriola pass, off Matia island; off Clayoquot sound; north of Marble island, Danger rocks near entrance to Houston Stewart channel, entrance to Flamingo harbour, in western portion and off Rose harbour in Houston Stewart channel, Massett harbour and outside Massett inlet (Fraser); San Diego, one location in each of the three sections of San Francisco bay; off Heceta head, Ore.; Channel rocks, Puget sound; east of Pillar bay, off Klashwan point, northeast of Reef island, Q.C.I.; Berg inlet, Ernest sound off Esterly island, off McArthur reef and Shingle island, in Sumner strait, Gardiner buoy off Admiralty island, Symonds point in Lynn canal, Alaska. Low tide to 240 fathoms.

Halecium washingtoni Nutting
Plate 23, Fig. 125

Halecium geniculatum NUTTING, Hyd. of Alaska and Puget sound, 1899, p. 744.

Halecium washingtoni NUTTING, Am. Nat., 1901, p. 780.

FRASER, West Coast Hyd., 1911, p. 50.

Vancouver Island Hyd., 1914, p. 170.

Hyd. from gulf of Alaska, 1914, p. 219.

Hyd. from west coast of V.I., 1935, p. 144.

Hyd. Distr. in vicinity of Q.C.I., 1936, p. 125.

Trophosome.—Stem and larger branches fascicled, but not so stout as some of the other fascicled stems of the genus; primary branches irregular, secondary branches less so, with an alternate arrangement; internodes long, annulated proximally, with at least two, often more, annulations, these usually oblique; the ultimate branches are bent at the nodes to give a zigzag appearance; pedicels usually annulated at the base; hydrophores with flaring rim; hydranths large.

Gonosome.—"Gonangia borne singly in the axils of the branches, regularly ovoid in one view, barnacle-shaped in the other; aperture large, terminal. The appearance of some of them would indicate the possible presence of an acrocyst at a later stage of development" (Nutting).

Distribution.—Puget sound (Nutting); San Diego (Torrey); Dodds narrows, San Juan archipelago; Clarke rock, Nanoose bay, north of Gabriola island, Northumberland channel, Gabriola reefs, off Matia island, Friday Harbor; gulf of Alaska; off Clayoquot sound, south of Flores island; off Rose harbour and in the western portion of Houston Stewart channel (Fraser); San Diego, off Southampton light in San Francisco bay; east of Pillar bay, off Klashwan point, 16 miles N.E. of Reef island, Q.C.I.   7 to 50 fathoms.

Halecium wilsoni Calkins

Plate 23, Fig. 126

Halecium wilsoni CALKINS, Hyd. of Puget sound, 1899, p. 343.
    FRASER, West Coast Hyd., 1911, p. 49.
        Vancouver Island Hyd., 1914, p. 170.
    Hyd. from west coast of V.I., 1935, p. 144.
        Hyd. Distr. in vicinity of Q.C.I., 1936, p. 125.

Trophosome.—Stem fascicled but slender and delicate; main branches few, irregularly arranged; branchlets numerous, short and slender; the internodes are of a uniform length, 1-3 annulations at each node; hydrophore slightly flaring.

Gonosome.—Male and female gonophores similar in shape but with the female smaller than the male, disk-shaped, with the opening at the distal end; the blastostyle is distinctly branched; the sporosacs are extruded into an acrocyst; irregularly situated near the base of the main branches or branchlets.

Distribution.—Bremerton (Calkins); Bare island (Hartlaub); Ucluelet, San Juan archipelago; Clarke rock, Nanoose bay, north of Gabriola island, Northumberland channel, Gabriola reefs, off Matia island, Friday Harbor; off Sydney inlet, Estevan point, Catala island, bar off Indian village in Esperanza inlet; Houston Stewart channel at the western end and off Rose harbour, outside Massett inlet, Q.C.I. (Fraser); off Klashwan point, off Massett sound, Q.C.I.; off McArthur reef in Sumner strait, Alaska.   15 to 45 fathoms.

Genus OPHIODISSA

Trophosome.—As in the family; tentacular organs present.
Gonosome.—Gonophores producing fixed sporosacs.

KEY TO SPECIES

A.   Stem creeping; hydrophores almost sessile......*O. carchesium*

B. Stem erect; hydrophores pedicellate
    a. Stolon, stems and pedicels strongly annulated.*O. corrugata*
    b. Stolon and stem with few or no annulations....*O. gracilis*

Ophiodissa carchesium (Fraser)
Plate 23, Fig. 127

Ophiodes carchesium FRASER, Hyd. from gulf of Alaska, 1914, p.220.

Trophosome.—Sub-sessile hydrophores grow directly from a loosely reticulate stolon, which shows no sign of division into internodes. The hydrophores are large and more campanulate than is usual in the *Halecidae.* Below the hydrophore there is a sharp constriction which separates the hydrophore from the basal support, which is too short to be properly called a pedicel. The hydranth is large, much greater in diameter than the width of the hydrophore; tentacles 10-12. Tentacular organs appear at intervals along the stolon; the terminal bulb is about twice the diameter of the chitinous cap that surrounds the stalk of the organ.

Gonosome.—Unknown.

Distribution.—Off Trinity islands, gulf of Alaska, 50 fathoms (Fraser).

Ophiodissa corrugata (Fraser)
Plate 23, Fig. 128

Ophiodes corrugata FRASER, Hyd. from Q.C.I., 1936, p. 504.

Trophosome.—Individual zooids, or small colonies, grow from a much annulated stolon. The individual zooids have short pedicels, sometimes less than 0.3 mm., but no matter how short they are, they have at least one or two annulations or corrugations. When a colony appears, the stem is annulated like the stolon, but the diameter is much less. The individuals in the colony are irregularly arranged, and each one looks like the individual zooid that is attached to the stolon. The terminal portion gradually increases in diameter to a prominent flare.

The tentacular organs are relatively large, sparsely scattered over the stolon, and over the stem of the colonies; none was observed directly on the hydrophore pedicels; they are tubular, but flare slightly at the margin.

Gonosome.—Unknown.

Distribution.—Off Rose harbour in Houston Stewart channel, 30 fathoms (Fraser).

8

Ophiodissa gracilis (Fraser)

Plate 23, Fig. 129

Ophiodes gracilis FRASER, Vancouver Island Hyd., 1914, p. 171.
Hyd. Distr. in vicinity of Q.C.I., 1936, p. 125.

Trophosome.—Slender stems growing from a regular reticular stolon; the longer stems are very distinctly jointed and may be slightly branched. Besides these longer stems, there are numerous shorter ones, each of which serves as a pedicel for a hydrophore; they are not jointed but they are distinctly annulated; the margin of the hydrophore flares but little, often not at all.

Tentacular organs appear on the stolon as well as on the stem; they are very long and slender and are protected by a chitinous cap, similar to a nematophore of the *Plumularidae.*

Gonosome.—Unknown.

Distribution.—Rose spit, Q.C.I.; in tow net, near surface, western portion of Houston Stewart channel (Fraser).

Family Hebellidae

Trophosome.—Colony simple, creeping; hydranth with conical or dome-shaped proboscis; hydrotheca tubular; diaphragm present; no operculum.

Gonosome.—Gonangia separated, not collected in a mass.

Genus HEBELLA

Trophosome.—Colonies consisting of single hydranths attached by short pedicels to a stolon, which usually creeps over other hydroids. A distinct diaphragm is present in the hydrotheca.

Gonosome.—Gonophores producing free medusae.

Hebella pocillum (Hincks)

Plate 24, Fig. 130

Lafoea pocillum HINCKS, Br. Hyd. Zooph., 1868, p. 204.
Hebella pocillum FRASER, West Coast Hyd., 1911, p. 51.

Trophosome.—Hydrothecae growing singly from a stolon, supported on obliquely annulated pedicels; hydrothecae minute, bulging somewhat at the base, then narrowing slightly and expanding again towards the entire margin.

Gonosome.—Unknown.

Distribution.—Nunivak island, Alaska (Clark); Kadiak, Alaska (Nutting); Jamestown bay, Southampton light in San Francisco bay; Channel islands, Puget sound; Sitka, Alaska. Low tide to 9 fathoms.

Family Lafoeidae

Trophosome.—Hydrothecae tubular; margin entire, operculum absent; there is no diaphragm, except a very thin one in *Lictorella*, in the hydrotheca; hydranth with a conical proboscis.

Gonosome.—Gonangia forming a "Coppinia" mass, except in the genus *Acryptolaria*, where there are not the special, modified hydrothecae protecting the gonangial aggregation.

KEY TO GENERA

A. Hydrothecae directly attached to a reticular stolon.. *Filellum*
B. Hydrothecae attached to a fascicled stem
    a. No indication of a diaphragm in the hydrotheca
        a. Peripheral tubes do not extend to the end of mature stems or branches................*Acryptolaria*
        b. Peripheral tubes extend to the end of mature stems or branches
            1. Hydrothecae free or very slightly adherent.. *Lafoea*
            2. Hydrothecae largely adherent or immersed.......
            ....*Grammaria*
    b. Some indication of a thin diaphragm in the hydrotheca...
        ....*Lictorella*

Genus ACRYPTOLARIA

Trophosome.—Stem and proximal portions of branches consist of a central tube, giving rise to hydrothecae, surrounded by a series of peripheral tubes, not bearing hydrothecae. The peripheral tubes do not extend to the ends of the branches; the hydrothecae are bilaterally arranged, usually free from the tube, where the peripheral tubes are present and partly adnate when the central tube is not covered.

Gonosome.—The tubular gonangia are aggregated but they lack the protective hydrothecae, present in a typical "coppinia".

Acryptolaria pulchella (Allman)
Plate 24, Fig. 131

Cryptolaria pulchella ALLMAN, Challenger Hyd., XXIII, 1888, p. 40.
            FRASER, New and previously unreported Hyd., 1925, p. 172.
Acryptolaria pulchella NUTTING, Philippine Hyd., 1927, p. 210.

Trophosome.—Colony with stout stem, irregularly branched; primary branches again branched; secondary branches very regularly alternate; hydrothecae regularly alternate.

Gonosome.—Unknown.

Distribution.—Off Goat island, San Francisco bay.   10 fathoms (Fraser).

Genus FILELLUM

Trophosome.—Stem a slender stolon, growing over other hydroids, wormtubes, *etc.*; hydrothecae partly adherent, the free portion curved upward; no diaphragm in the hydrothecal cavity.

Gonosome.—A coppinia mass.

Filellum serpens (Hassall)
Plate 24, Fig. 132

Campanularia serpens HASSALL, Trans. Micr. Soc., 1852, p. 163.
Filellum serpens FRASER, West Coast Hyd., 1911, p. 50.
            Vancouver Island Hyd., 1914, p. 172.
            Hyd. from west coast of V.I., 1935, p. 144.
            Hyd. Distr. in vicinity of Q.C.I., 1936, p. 145.

Trophosome.—Stolon reticular, creeping over other hydroids, wormtubes, *etc.*; hydrothecae adherent for from one-half to two-thirds of their length; nearly the same size throughout, not annulated, but there may be some horizontal striae near the margin; margin not flaring.

Gonosome.—Coppinia mass rather compact, the gonangia not placed so close together as in other species; hydrothecal tubes long and slender.

Distribution.—Juneau, Alaska (Nutting); San Juan archipelago; Nawhitti bar, Hammond bay, Departure bay, Northumberland channel, Friday Harbor; off Long beach; entrance to Flamingo harbour, western portion of Houston Stewart channel and off Rose harbour (Fraser); San Diego, one location in the upper section of San Francisco bay; off Yaquina light, Ore.; off Frederick island, off

Sadler point, Q.C.I.; Juneau, off Easterly island in Ernest sound, off McArthur reef in Sumner strait, Alaska. Low tide to 87 fathoms.

Genus GRAMMARIA

Trophosome.—Stem fascicled, consisting of a hydrothecate, axial tube, surrounded by a certain number of peripheral, non-hydrothecate tubes; hydrothecae partly adherent; no diaphragm in hydrothecal cavity.

Gonosome.—A coppinia mass.

KEY TO SPECIES

A. Large portion of the hydrotheca free from the stem..*G. abietina*
B. Hydrothecae almost wholly immersed..........*G. immersa*

Grammaria abietina (Sars)
Plate 24, Fig. 133

Campanularia abietina SARS, Nyt. Mag. for Naturvid., 1851, p. 139.
Grammaria abietina FRASER, Vancouver Island Hyd., 1914, p. 173.

Trophosome.—Stem very stout, irregularly branched, branches constricted at the base, resembling a main stem in all particulars; a large portion of the hydrotheca free, the free portion being directed outward; orifice nearly circular; margin vertical.

Gonosome.—"Coppinia generally of an irregular oval form. All the tubes extending radially from it, bend at a certain distance from the surface in all directions, thus forming a network lying like a capsule outside the cluster of gonangia" (Bonnevie).

Distribution.—Swiftsure shoal, Departure bay, Northumberland channel, Friday Harbor (Fraser); west end of Houston Stewart channel; off McArthur reef in Sumner strait; point Gardiner buoy off Admiralty island, Alaska. 15 to 40 fathoms.

Grammaria immersa Nutting
Plate 24, Fig. 124

Grammaria immersa NUTTING, Hyd. of the Harriman Exp., 1901, p. 176.

FRASER, West Coast Hyd., 1911, p. 50.

Vancouver Island Hyd., 1914, p. 174.

Hyd. Distr. in vicinity of Q.C.I., 1936, p. 125.

Trophosome.—Stem stout but not so stout as that of *G. abietina*; branches given off at regular intervals, forming nearly a right angle with the stem, constricted at the base; hydrothecae so much immersed that only a short portion of the distal extremity shows; the free portion is nearly at right angles to the stem and the margin is vertical.

Gonosome.—Coppinia mass almost globular but slightly elongated in the direction of the stem; one mass measured 6 mm. long and 4 mm. broad, and that relation holds very nearly for all the other masses examined; the hydrothecae are long, slender, and very numerous, much similar to those in some species of *Lafoea.*

Distribution.—St. Paul's harbour and Kadiak, Alaska (Nutting); Bering sea, northwest of St. Lawrence island (Jäderholm); Dodds narrows; off Lasqueti island, Northumberland channel, Gabriola reefs; western portion of Houston Stewart channel (Fraser); Nawhitti bar, off Sadler point, Q.C.I.; Kadiak, Alaska. 8 to 30 fathoms.

Genus LAFOEA

Trophosome.—Mature stems strongly fascicled and erect, young stems may be creeping; hydrothecae, with but few exceptions, entirely free from the stem; no diaphragm in the hydrothecal cavity.

Gonosome.—A coppinia mass.

KEY TO SPECIES

A. Hydrothecae growing from creeping stolons
    a. Stolons closely interwoven................*L. adhaerens*
    b. Stolons forming a loose network.............*L. adnata*
B. Stems erect and strongly fascicled
    a. Hydrothecae sessile, sometimes slightly adherent at the base.................................*L. dumosa*
    b. Hydrothecae pedicellate
        a. Pedicel making an angle of less than 45° with the stem
        ....*L. gracillima*
        b. Pedicel making an angle of more than 45° with the stem
        ....*L. fruticosa*

Lafoea adhaerens Nutting
Plate 24, Fig. 135

Lafoea adhaerens NUTTING, Hyd. of the Harriman Exp., 1901, p. 178.

FRASER, West Coast Hyd., 1911, p. 51.

Trophosome.—"Colony forming an encrusting mass of adherent rootstocks, disposed both longitudinally and transversely, over colonies of other hydroids, the tubes of the rootstocks interwoven much like the threads of a fabric. Hydrothecae sessile, tubular, often more or less curved; aperture facing upward, entire; margin slightly expanded. The hydrothecae are very irregularly disposed, being much more crowded in some places than others" (Nutting).

Gonosome.—"The Coppinia mass is much like that of *Lafoea dumosa*, being composed of closely packed gonangia interspersed with long, tubular, variously curved, modified hydrothecae. The gonangia are flask-shaped, with a tubular neck and small aperture. Each gonangium apparently contains a single ovum" (Nutting).

Distribution.—Kadiak, Alaska (Nutting).

Lafoea adnata Fraser
Plate 25, Fig. 136

Lafoea adnata FRASER, New and previously unreported Hyd., 1925, p. 170.

Trophosome.—Hydrothecae attached to a reticulate stolon, with no definite pedicels, although there is a constriction at the base of each; in nearly all cases, the wall of the hydrotheca is attached to the stolon for a part of its length but the proportion varies very materially; except for the tapering to the constriction, the hydrotheca is of uniform diameter; there may or may not be a slight flare at the margin; the hydrotheca is always curved but the amount of the curvature varies; in extreme cases, the distal extremity is at right angles to the proximal; there is no diaphragm present; hydranth with 8 tentacles.

Gonosome.—Unknown.

Distribution.—Near Farallon islands in 33-35 fathoms (Fraser).

Lafoea dumosa (Fleming)
Plate 25, Fig. 137

Sertularia dumosa FLEMING, Phil. Jour., II, 1828, p. 83.
Lafoea dumosa NUTTING, Hyd. from Alaska and Puget sound, 1899, p. 741.
FRASER, West Coast Hyd., 1911, p. 51.
Vancouver Island Hyd., 1914, p. 174.
Hyd. from west coast of V.I., 1935, p. 144.
Hyd. Distr. in vicinity of Q.C.I., 1936, p. 125.

Trophosome.—Mature stem strongly fascicled, erect, coarse, much branched, branches also coarse; young stem either erect or creeping over other hydroids; hydrothecae sessile but usually free from the stem; occasionally towards the distal part of the stem the hydrothecae are adherent; even when the base is free, the proximal portion often passes up in the same direction as the stem, the distal portion curving outward.

Gonosome.—The gonangia of the coppinia mass, as seen from the surface, are hexagonal with a projection containing the orifice at the centre; the elongated hydrothecae come out at intervals among them.

Distribution.—Port Etches, Alaska (Clark); Puget sound, Berg inlet, Orca, Alaska; Port Orchard, Puget sound (Torrey); Banks island, Departure bay, Dodds narrows, Ucluelet, Port Renfrew, San Juan archipelago; found almost everywhere that collections have been made from Queen Charlotte islands to Puget sound; off Sydney inlet; entrance to Flamingo harbour, in Houston Stewart channel at the western end and off Rose harbour, Massett harbour, outside Massett sound (Fraser); San Pedro, several locations just outside and just inside the Golden Gate; Port Orchard, Puget sound; Nawhitti bar, off Tian head, Rennell sound, off Sadler point, east of Sand spit, Q.C.I.; off McArthur reef in Sumner strait, Alaska. Low tide to 60 fathoms.

Lafoea fruticosa Sars
Plate 25, Fig. 138

Lafoea fruticosa SARS, Norske Hydroider, 1862, p. 30.
FRASER, West Coast Hyd., 1911, p. 52.
Vancouver Island Hyd., 1914, p. 175.
Hyd. Distr. in vicinity of Q.C.I., 1936, p. 125.

Trophosome.—Stem fascicled, with many large branches regularly arranged, sometimes all on one side of the stem; pedicels long, with 3 or 4 twists, passing out at an angle of more than 45° with the stem; hydrotheca large, with the lower side more nearly in line with the pedicel than the upper side; the margin is at right angles to the wall of the hydrotheca.

Gonosome.—"Coppinia with small irregular facets and tubes that are very long and thin and curved in a spiral like a watch-spring" (Bonnevie).

Distribution.—Shumagin islands to Kyoka island, Alaska (Clark); Puget sound, Juneau, Berg inlet, Orca, Alaska (Nutting), Bering sea (Jäderholm); off cape Edenshaw, Swiftsure shoal, north of Gabriola island, Gabriola reefs; near Trinity islands, gulf of Alaska; north of Marble island (Fraser); near Shag rock in the lower section of San Francisco bay; off Klashwan point, off Massett sound, Q.C.I.; off cape James, Hope island; Juneau, Mill creek, off Shingle island in Sumner strait, Gardiner buoy off Admiralty island, Symonds point in Lynn canal, Alaska. 5 to 240 fathoms.

Lafoea gracillima (Alder)
Plate 25, Fig. 139

Campanularia gracillima ALDER, Trans. Tynes. F.C., 1857, p. 39.
Lafoea gracillima FRASER, West Coast Hyd., 1911, p. 52.
      Vancouver Island Hyd., 1914, p. 175.
      Some Alaskan Hyd., 1914, p. 220.
      Hyd. from west coast of V.I., 1936, p. 144.
      Hyd. Distr. in vicinity of Q.C.I., 1936, p. 125.

Trophosome.—Stem fascicled, very much branched, but without the distinct main stem present in *L. dumosa* and *L. fruticosa*; the young colonies often are creeping; hydrothecae long, tubular, somewhat curved, with the convex side uppermost, smaller than those of the other species but varying much in size; pedicels with one or more slight twists, coming out from the stem at an angle less than 45°.

Gonosome.—The whole coppinia mass resembles that of *L. dumosa* but the separate gonangia are more regularly hexagonal when viewed from the surface; they are more nearly circular and are not so regularly arranged; the hydrothecal tubes are more slender and perhaps longer.

Distribution.—Sitka harbour to Shumagin islands (Clark); Puget sound, Juneau, Berg inlet, Orca, Alaska (Nutting); San Pedro, Puget sound (Torrey); Bare island (Hartlaub); Departure bay, Dodds narrows, Ucluelet, Port Renfrew, San Juan archipelago; found almost everywhere where collections have been made from Queen Charlotte islands to Puget sound; Trinity island, gulf of Alaska; off Clayoquot sound; north of Marble island, west of Horn island in Tasoo harbour, western portion of Houston Stewart channel, outside Massett inlet (Fraser); three locations in the upper section and one in the lower section in San Francisco bay, outside the Golden Gate, off Bonita point; western portion of Goose island halibut grounds, off Virago sound, off Massett inlet, off cape James, Hope island; Queen Charlotte sound; Juneau, Berg inlet, Kadiak, off Easterly island in Ernest sound, Mill creek, off McArthur reef in Sumner strait, off Shingle island in Sumner strait, point Gardiner buoy off Admiralty island, Alaska.   3 to 240 fathoms.

Genus **LICTORELLA**

Trophosome.—Stem fascicled, with ultimate branches monosiphonic and bilateral; hydrothecae never sessile; delicate structure serving as a diaphragm in the hydrothecal cavity; nematocysts may be present on the branch at the base of the hydrothecae.

Gonosome.—"Gonangia aggregated, with curious protuberant shoulders on one or two sides at the distal end.   These are horn-like processes which may curve upward or downward, or be directed straight outward, according to the species" (Nutting).

KEY TO SPECIES

A.   Hydrotheca with one side straight and the other slightly convex or straight..............................*L. carolina*
B.   Hydrotheca with a distinct short curve near the margin......
....*L. cervicornis*

Lictorella carolina Fraser
Plate 26, Fig. 140

Lictorella carolina FRASER, West Coast Hyd., 1911, p. 53.
Vancouver Island Hyd., 1914, p. 176.
Some Alaskan Hyd., 1914, p. 221.

Trophosome.—Stem and larger branches fascicled; the tubes gradually decrease in number as the branches get farther from the base; branching has a dichotomous appearance; largest colony 35 mm.; the ultimate branches are divided into internodes of almost equal length by deep constrictions; from each internode, nearly midway between the nodes, a single hydrotheca is given off.   These hydrothecae are alternate on successive internodes but are all in the same plane; at the origin of each of these there is a distinct shoulder on the branch, which is divided by a deep constriction from the base of the hydrotheca; on this shoulder there are two nematocysts that are deeply cup-shaped and are supported by a two-ringed pedicel. The hydrothecae widen gradually and symmetrically until the diaphragm is reached; from this point the under surface passes out almost in a straight line, while the upper surface is convex for some distance, after which it passes out parallel to the lower side; the margin is but slightly flaring; it is commonly duplicated.

Gonosome.—Unknown.

Distribution.—San Juan archipelago; Trinity islands, gulf of Alaska (Fraser).   50 fathoms.

Lictorella cervicornis Nutting
Plate 26, Fig. 141

Lictorella cervicornis NUTTING, Hawaiian Hyd., 1905, p. 934.
                FRASER, Monobrachium parasitum, *etc.*, 1918,
                p. 134.

Trophosome.—Colony with a continuous main stem, reaching a height of nearly 50 mm. in largest specimen; branches given off alternately, at a wide angle with the stem; most of them, in some cases, all of them, unbranched; main stem and larger branches fascicled; the processes that support the hydrothecae are regularly arranged on the stem and branches; hydrothecae deep, almost tubular; the proximal end narrows into the pedicel and the distal end broadens slightly; just proximal to the margin there is a distinct but short curve in the tube, with the abcauline side convex, aperture round; margin entire; pedicel short, making a distinct joint with the process from the stem or branch.   Hydrothecae given off from the fascicled portion of the stem, as well as those in the axils of the branches, as a rule, have longer pedicels than those on the monosiphonic branches.   There is a nematocyst present in the axil of each pedicel process.

Gonosome.—"Gonangia forming a coppinia mass on the main stem, the distal ends being the broader on account of the opposite shoulders, which are quite conspicuous and end in round apertures. Midway between these shoulders there is a short neck ending in a third aperture.   The individual gonangia are borne on short branchlets, which continue beyond them, arching over each gonangium so as to form a protecting network of such branches over the aggregated gonangia.   This structure seems to resemble quite closely the phylactogonia found in certain genera of plumularian hydroids" (Nutting).

Distribution.—North of Gabriola island, north of Snake island (Fraser); Kaison bank, Moresby island; Mill creek, Alaska.   30 to 110 fathoms.

Family Synthecidae

Trophosome.—Stem branched; hydrothecae sessile and, to some extent, adnate to the stem or branch; margin entire; no operculum.

Gonosome.—Gonophores produce fixed sporosacs.

Genus SYNTHECIUM

Trophosome.—Stem divided into regular internodes; branches opposite or alternate, also divided into regular internodes; hydrothecae opposite or alternate; adnate to the stem or branch; margin entire, circular or saddle-shaped; no operculum.

Gonosome.—Gonophores arising from the interior of the hydrothecae, taking the place of hydranths.

Synthecium cylindricum (Bale)
Plate 26, Fig. 142

Sertularella cylindrica BALE, Proc. Linn. Soc., N.S.W., (2), 3, 1888, p. 765.

Sertularella halecina TORREY, Hyd. of the Pacific coast, 1902, p. 61.

Synthecium cylindricum NUTTING, Am. Hyd., II, 1904, p. 136.

FRASER, West Coast Hyd., 1911, p. 74.

Trophosome.—Stems 25 mm. or less in length, arising from a creeping stolon, unbranched or very slightly branched; branches similar to the stem, given off at a wide angle, a hydrotheca in the axil; stem and branches divided into regular internodes, each bearing a hydrotheca, but sometimes the nodes are faint or appear to be absent; hydrothecae regularly alternate, rather distant, the oblique base being adnate, the remainder free; cylindrical, slightly curved; margin wide, circular, flaring; operculum lacking.

Gonosome.—Gonangia arising from hydrothecae in place of hydranths, with stout, slightly sinuous pedicels of considerable length; almost spherical. A whole series of fine wavy lines pass from the margin of the gonophore to the gonangium.

Distribution.—San Diego, 5-12 fathoms (Torrey); San Diego, Farallone islands, Southampton light, Goat island and No. 7 beacon, in the lower section of San Francisco bay.

Family Sertularidae

Trophosome.—Hydrothecae sessile, usually arranged on both sides of the stem and branches, and more or less adnate to them; operculum present.

Gonosome.—Gonophores producing fixed sporosacs.

KEY TO GENERA

A. Hydrothecae all on one side of the branches, their distal end alternating to right and left. . . . .*Hydrallmania*
B. Hydrothecae arranged in two longitudinal rows
    a. Hydrothecae in opposite pairs. . . . .*Sertularia*
    b. Hydrothecae alternate
        a. Operculum of one adcauline flap
           1. Hydrothecal aperture small, body flask-shaped. . .
           . . . .*Abietinaria*
           2. Hydrothecal aperture large, body not flask-shaped
           . . . .*Diphasia*
        b. Operculum abcauline,* or with more than one flap
           1. Operculum of one or two flaps. . . . .*Thuiaria*
           2. Operculum of 3 or 4 flaps. . . . .*Sertularella*
C. Hydrothecae arranged on all sides of the branches
    a. Operculum usually of one flap. . . . .*Selaginopsis*
    b. Operculum of several flaps. . . . .*Dictyocladium*

Genus **ABIETINARIA**

Trophosome.—Hydrothecae alternate, flask-shaped; aperture small; operculum with a single adcauline flap.

Gonosome.—Gonangia simple, without spines or internal marsupium.

KEY TO SPECIES

A. Stem stout
    a. Cauline nodes indicated by a double annulus. .*A. annulata*
    b. Cauline nodes with a single annulus
        a. Gonangia with strong longitudinal crests
           1. Primary branches usually unbranched.*A. amphora*
           2. Primary branches much branched. . . . . .*A. turgida*
        b. Gonangia smooth or annulated transversely
           1. Hydrothecae almost wholly immersed. .*A. gigantea*
           2. One-fourth to one-third of the hydrotheca free
              i. Margin vertical or nearly so. . . . .*A. rigida*
              ii. Margin nearly horizontal. . . . . . .*A. variabilis*
           3. Much more than one-third of the hydrotheca free
              i. Hydrotheca with distinct neck
                  I. Branches making a wide angle with the stem; hydrotheca large. . .*A. abietina*

---

*Except in **Thuiaria thuiaroides**, an intergrading form, where the flap is adcauline.

Abietinaria abietina (Linnaeus)
Plate 27, Fig. 143

Sertularia abietina LINNAEUS, Syst. Nat., 1758, p. 808.
Abietinaria abietina NUTTING, Am. Hyd., II, 1904, p. 114.
               FRASER, West Coast Hyd., 1911, p. 57.
               Vancouver Island Hyd., 1914, p. 178.
               Hyd. of west coast of V.I., 1935, p. 144.
               Hyd. Distr. in vicinity of Q.C.I., 1936,
               p. 125.

Trophosome.—Main stem stout, straight or slightly flexuous, divided into regular internodes; primary branches large, pinnately arranged, with three hydrothecae between two successive branches on the same side of the stem. The primary branches are occasionally branched; they are divided into regular internodes. Hydrothecae large, occasionally nearly opposite, large at the base, narrowing above to form a distinct neck and expanding again slightly to the round, smooth margin, which is horizontal but oblique as compared with the axis of the hydrotheca; a large portion of the hydrotheca, sometimes more than half, is free from the stem.

Gonosome.—Gonangia borne on the upper sides of branches; oval with a short collar and a wide aperture; smooth or very slightly annulated.

Distribution.—Alaska, Bering sea, several *Albatross* stations off Washington, British Columbia, and Alaska (Nutting); San Juan archipelago, Ucluelet, Dodds narrows, Departure bay, Banks island; very abundant from Queen Charlotte islands to Puget sound; off Long beach, off Sydney inlet, Nootka island, Bajo reef; Houston Stewart channel at the western end and off Rose harbour (Fraser); San Diego, Lime point, San Francisco bay; Nawhitti bar, western portion of Goose island halibut grounds, off Sadler point, 28 miles northeast of Scudder point, off Roller bay and off cape James, Hope island. 8 to 80 fathoms.

Abietinaria alexanderi Nutting
Plate 27, Fig. 144

Abietinaria alexanderi NUTTING, Am. Hyd., II, 1904, p. 120.

FRASER, West Coast Hyd., 1911, p. 58.

Trophosome.—Colony large; main stem straight, divided into quite regular internodes, with a branch and two hydrothecae on one side and one hydrotheca on the other, the branches alternating; the branches are divided irregularly into internodes with several hydrothecae, or are without divisions. Hydrothecae sub-alternate; the basal portion tumid, the distal portion tubular; margin oval, entire or sinuous; operculum a single adcauline flap.

Gonosome.—"Gonangia borne mostly in rows on upper sides of distal branches, small, ovoid, without neck; aperture obscurely polygonal, marked by four or five fine, dark, meridional lines, giving the effect of radial canals on sessile medusae" (Nutting).

Distribution.—South of Unimak pass, south of northwestern Aleutians, Alaska, 55-56 fathoms (Nutting).

Abietinaria amphora Nutting
Plate 27, Fig. 145

Abietinaria amphora NUTTING, Am. Hyd., II, 1904, p. 119.

FRASER, West Coast Hyd., 1911, p. 58.

Vancouver Island Hyd., 1914, p. 179.

Hyd. from west coast of V.I., 1935, p. 144.

Hyd. Distr. in vicinity of Q.C.I., 1936, p. 125.

Trophosome.—Stems straight, with branches very regularly arranged, with three hydrothecae between two successive branches on the same side of the stem; the portion near the base that is free of branches is also devoid of hydrothecae; this portion is deeply annulated. The primary branches do not branch again, and seldom have noticeable nodes; hydrothecae nearly opposite, swollen at the base and narrowing to a neck like those of *A. abietina*, but they are not so large as these; about half free; margin entire.

Gonosome.—Gonangia borne on the front of the stem and proximal portions of the branches, usually crowded together in two rows; very large, oval, with long neck and round, terminal aperture, provided with 4 or 5 strong crests running longitudinally from the neck to the base.

Distribution.—Off Unimak pass, entrance to strait of Fuca, Whidby island (Nutting); Port Renfrew, Ucluelet, Dodds narrows; Northumberland channel, Pylades channel, Gabriola pass, Nawhitti bar, Friday Harbor, San Juan archipelago; bar off Indian village in Esperanza inlet; Danger rock, off Rose harbour, Flat rock and western portion of Houston Stewart channel, entrance to Big bay (Fraser); outside of Golden Gate, one location in the upper section and two in the lower section of San Francisco bay; off Heceta head, Ore.; Nawhitti bar, off Sadler point, off Massett inlet, 5 miles east of Sand spit, off cape James, Hope island; Yakutat, off McArthur reef in Sumner strait, Alaska.   Low tide to 171 fathoms.

Abietinaria anguina (Trask)

Plate 27, Fig. 146

Sertularia anguina TRASK, Proc. Cal. Acad. Sc., 1857, p. 112.
Abietinaria anguina FRASER, West Coast Hyd., 1911, p. 58.
                    Vancouver Island Hyd., 1914, p. 179.
                    Hyd. from west coast of V.I., 1935,
                      p. 145.
                    Hyd. Distr. in vicinity of Q.C.I., 1936,
                      p. 125.

Trophosome.—Stem rather slender, strongly geniculate above the lowest branches, straight below these and annulated; branches regularly pinnate, not branched again.   One hydrotheca in the axil of each branch and one next it on the same side; hydrothecae nearly opposite, of the *abietina* type but small; margin scarcely even, rather shovel-shaped.

Gonosome.—Gonangia growing from the upper side of the

branches, top-shaped, with a distinct collar, annulated, annulations near together proximally but farther apart distally.

Distribution.—San Diego (Hemphill); Monterey bay (Anderson); San Francisco (Trask); Vancouver island (Dawson); Tledis village, near Susk, B.C. (Nutting); Port Renfrew, Ucluelet; Hope island, Nawhitti bar; Estevan point, near Maquinna point, Bajo reef; Danger rocks, Flat rock and western portion of Houston Stewart channel, Massett harbour (Fraser); San Diego, San Pedro, Pacific Grove, Aumentos rocks, Cabrillo point, Monterey, point Lobos, Mile rock and Bonita point, outside Golden Gate, several locations in each of the three sections of San Francisco bay, point Reyes peninsula, Cal.; off Heceta head, Ore.; western portion of Goose island halibut grounds, off Frederick island, 5 miles east of Sand spit.   Low tide to 35 fathoms.

Abietinaria annulata (Kirchenpauer)
Plate 28, Fig. 147

Thuiaria annulata KIRCHENPAUER, Nordische Gattungen, 1884, p. 26.
Abietinaria annulata, NUTTING, Am. Hyd., II, 1904, p. 122.
FRASER, West Coast Hyd., 1911, p. 60.

Trophosome.—"Colony about 4 inches high.   Main stem and branches exceedingly thick and woody, black in color; the main branches spring from near base of the stem.   Stem and main branches straight, divided into irregular internodes, each of which bears several closely-approximated and upward-directed branches, each with an axillary hydrotheca; internodes with wide, shallow and equidistant annulations, which, in a general way, correspond in number to the hydrothecae on each side of the internode.   Branches divided into irregular and distant internodes, each with several hydrothecae on each side.   Hydrothecae sub-opposite, very closely approximated, short, stout, tubular, with slightly constricted distal ends; margin even, aperture nearly round, and either horizontal or slightly inclined towards the stem.   Operculum of one flap, attached to adcauline side of margin" (Nutting).

Gonosome.—Unknown.

Distribution.—South of Unimak pass, Alaska, 36 fathoms (Nutting).

9

Abietinaria costata (Nutting)

Plate 28, Fig. 148

Thuiaria costata NUTTING, Hyd. of the Harriman Exp., 1901, p. 187.
Abietinaria costata NUTTING, Am. Hyd., II, 1904, p. 122.

FRASER, West Coast Hyd., 1911, p. 60.

Hyd. Distr. in vicinity of Q.C.I., 1936, p. 125.

Trophosome.—Main stem rather slender, straight; the proximal part unbranched, divided into regular internodes by oblique nodes, each internode with two nearly opposite hydrothecae; the internodes of the more distal part of the stem bear a branch and two hydrothecae on one side and one hydrotheca on the other. The branches are dichotomously branched, with nodes irregularly placed, several hydrothecae to each internode, in nearly opposite pairs. Hydrothecae small, shaped like those of *A. abietina*, those on the two sides not quite in the same plane; margin circular, entire. A chitinous thickening projects backward from the inner, lower angle of each hydrotheca.

Gonosome.—Gonangia borne on both faces of the stem mainly, but sometimes on the proximal portion of the branches; almost sessile, oblong or obovate, with short, tubular neck and entire margin. There are five strongly marked longitudinal ridges. They bear some resemblance to those of *A. amphora*.

Distribution.—Yakutat, Alaska (Nutting); Danger rocks near eastern entrance to Houston Stewart channel, low tide (Fraser); Yakutat, Alaska.

Abietinaria filicula (Ellis and Solander)

Plate 28, Fig. 149

Sertularia filicula ELLIS and SOLANDER, Nat. Hist. Zooph., 1786, p. 57.

Abietinaria filicula NUTTING, Am. Hyd., II, 1904, p. 117.

FRASER, West Coast Hyd., 1911, p. 60.

Vancouver Island Hyd., 1914, p. 180.

Hyd. from west coast of V.I., 1935, p. 145.

Hyd. Distr. in vicinity of Q.C.I., 1936, p. 125.

Trophosome.—Stem slender, straight in unbranched portion slightly flexuous in branched portion; branches regularly pinnate, often branched again and sometimes more than once. The main stem is divided into regular internodes but the branches have few, if any, nodes and these irregularly placed; hydrothecae nearly opposite, shaped like those of *A. abietina* but smaller.

Gonosome.—Gonangia oval, tapering into a narrow neck above and into a pedicel, which is somewhat curved, below; surface smooth.

Distribution.—Alaska, at entrance to strait of Juan de Fuca (Nutting); San Juan archipelago, Victoria, Dodds narrows; Alert bay, Departure bay, Northumberland channel, Gabriola pass, Friday Harbor; south of Flores island; off Rose harbour and in western portion of Houston Stewart channel (Fraser); off Alcatraz island and off Angel island in San Francisco bay; Newport, Ore.; west of Nawhitti bar, western portion of Goose island halibut grounds, off Frederick island. 10 to 35 fathoms.

Abietinaria gigantea (Clark)
Plate 28, Fig. 150

Thuiaria gigantea CLARK, Alaskan Hyd., 1876, p. 230.
Abietinaria gigantea NUTTING, Am. Hyd., II, 1904, p. 123.
FRASER, West Coast Hyd., 1911, p. 60.
Vancouver Island Hyd., 1914, p. 180.
Hyd. Distr. in vicinity of Q.C.I., 1936, p. 125.

Trophosome.—Stem stout, with but few branches, these irregularly arranged and constricted at the base, similar in all respects to the main stem; nodes absent; hydrothecae arranged alternately with but a short space of the stem between each pair, almost wholly immersed; regularly curved and almost the same size throughout; margin nearly vertical.

Gonosome.—Gonangia arranged in two rows on the surface of the stem, oval, with a distinct pedicel, smooth or slightly wrinkled; distal and truncate, sometimes oblique; without collar.

Distribution.—Alaskan shores, Aleutian islands, Bering sea, Hagmeister island, Akutan pass, Kyska harbour, Orca, Kadiak, Belkovsky, southeast of Shumagin islands, off Saint Paul island, Bering sea, all in Alaska, 26-40 fathoms (Nutting); off Rose harbour and in western portion of Houston Stewart channel (Fraser); San Pablo bay and off Shag rock in San Francisco bay; Kadiak, Alaska. 5 to 30 fathoms.

Abietinaria gracilis Nutting
Plate 29, Fig. 151

Abietinaria gracilis NUTTING, Am. Hyd., II, 1904, p. 120.
          FRASER, West Coast Hyd., 1911, p. 61.
                 Vancouver Island Hyd., 1914, p. 181.
                 Some Alaskan Hyd., 1914, p. 221.

Trophosome.—Stem straggling; the portion that is branched, somewhat geniculate; branching pinnate, not regular, branches varying in length so that outline of the colony is irregular; hydrothecae alternate or nearly opposite, distant, basal portion enlarged, narrowing gradually to a slender neck which expands to form the margin, this turned upward so as to be transverse to the axis of the branch. The whole hydrotheca is longer and more slender than others of this type.

Gonosome.—Gonangia borne in rows on upper sides of branches; oval, narrowing above to form a short neck and below to form a pedicel; longitudinal crests present, distinct but not very much raised.

Distribution.—At the entrance to the strait of Juan de Fuca, south of the Aleutian islands, 40-283 fathoms (Nutting); north of Gabriola island, Northumberland channel; Trinity islands, gulf of Alaska (Fraser).

Abietinaria greenei (Murray)
Plate 29, Fig. 152

Sertularia greenei MURRAY, Ann. and Mag. N.H., (3), V, 1860,
          p. 504.
Abietinaria greenei NUTTING, Am. Hyd., II, 1904, p. 121.
          FRASER, West Coast Hyd., 1911, p. 61.
                 Vancouver Island Hyd., 1914, p. 181.
                 Hyd. as a food supply, 1933, p. 220.
                 Hyd. from west coast of V.I., 1935, p.
                 145.
                 Hyd. Distr. in vicinity of Q.C.I., 1936,
                 p. 125.

Trophosome.—Colony consisting of a dense cluster of slender, erect stems, which branch irregularly, the branches running in the same general direction as the stem, and these as well as the main stem, often branching dichotomously; the angle between the branch and the stem, even when the branching is dichotomous, is small;

there is a constriction of the branch at the base; hydrotheca alternate or nearly opposite, more erect than in most of the other species of *Abietinaria*; the proximal half is of the same diameter throughout; above this the hydrotheca narrows to form a neck which expands slightly to form the margin; two teeth are present on the abcauline side of the margin; these are usually very strongly marked but in some cases are scarcely noticeable.

Gonosome.—Gonangia borne in long rows on the front of branches, obconical, with a small curved pedicel and a distal tapering collar; aperture circular, surface regularly, but not deeply, corrugated. The sporosacs pass out of the gonangium into an acrocyst.

Distribution.—Tomales point, Monterey, Punta Reyes, San Francisco, Port Renfrew (Nutting); San Juan archipelago, Dodds narrows, Departure bay, Port Renfrew, Ucluelet; Northumberland channel, Pylades channel, Gabriola pass, Nawhitti bar, Friday Harbor, Copalis, Port Grenville; in the stomach of *Charitonetta albeola* in Barkley sound; Nootka island, Bajo reef; Danger rocks, Flat rock and in the western portion of Houston Stewart channel, entrance to Flamingo harbour, entrance to Big bay; Massett harbour (Fraser); off Mile rock, outside the Golden Gate, Alcatraz island and San Pablo point in San Francisco bay; Newport, Ore.; Nawhitti bar, off Frederick island. Low tide to 20 fathoms.

Abietinaria inconstans (Clark)

Plate 29, Fig. 153

Sertularia inconstans CLARK, Alaskan Hyd., 1876, p. 222.
Abietinaria inconstans NUTTING, Am. Hyd., II, 1904, p. 116.
                    FRASER, West Coast Hyd., 1911, p. 61.
                    Hyd. as a food supply, 1933, p. 260.

Trophosome.—Stem short but stiff and coarse; proximal unbranched portion with numerous distinctly marked nodes; nodes on the branched portion oblique, with two hydrothecae and a branch on one side of the internode, and one hydrotheca on the other; branches make a very acute angle with the stem; they have two or more marked annulations at the base; the internodes are not of equal length but are always short; hydrothecae nearly opposite, resembling somewhat those of *A. abietina*, although they are much smaller. The proximal portion is more swollen relatively than in that species; usually a chitinous projection extends backward at the lower, inner angle.

Gonosome.—"The gonangia show the greatest amount of variation of any species that I know of; it is impossible to describe their form, for there is not one of them that seems to agree with any other. Sessile, large, orifice terminal, small, discoidal, outline very irregular, tapering usually at the base, borne in two rows on distal portion of the main stem" (Clark).

Distribution.—Unalaska beach (Clark); in the stomach of *Somateria spectabilis* in Hooper bay, Alaska (Fraser).

Abietinaria pacifica Stechow
Plate 29, Fig. 154

Abietinaria pacifica STECHOW, Hyd. Fauna des Mittelsmeeres, *etc.*, 1923, p. 197.

Trophosome.—(After Stechow, who described the species from a fragment 6 mm. long, found on the back of a crab). Stem monosiphonic, almost unjointed. There were five definitely alternate branches. Between two succeeding branches on the same side of the stem there are three widely separated hydrothecae. Except where the branch joins the process from the stem there are no joints in the branch. The longest branch, with 19 hydrothecae, was 7 mm.; a hydrotheca is present in the axile; hydrothecae regularly alternating, not closely arranged (the base of each hydrotheca somewhat above the margin of the hydrotheca on the other side), attached to the stem for three-fourths of their length; the side next the stem is convex, the other slightly concave. The hydrotheca narrows to the margin but does not form a definite neck; the margin is at right angles to the branch, directed upward; operculum very distinct, with one adcauline flap. A pointed chitinous thickening of the inner, lower angle of each hydrotheca is directed backward. Thickness of stem .35 mm.; thickness of branch .21 mm.; length of hydrotheca .35 mm.; breadth at the base .12 mm.; greatest breadth .18 mm.; breadth at margin .095 mm.

Gonosome.—Lacking.

Distribution.—Pacific Grove (Stechow).

Abietinaria rigida Fraser
Plate 29, Fig. 155

Abietinaria rigida FRASER, West Coast Hyd., 1911, p. 61.

Vancouver Island Hyd., 1914, p. 182.

Hyd. Distr. in vicinity of Q.C.I., 1936, p. 125.

Trophosome.—Main stem stout, rigid, straight or but slightly flexuous, with annulations at the base but very few nodes in the remainder of the stem; stems are found over 50 mm. in length, that are entirely unbranched; branches with regular pinnate arrangement, making a wide angle with the stem and constricted at the base; like the stem, they are rigid and brittle so that a colony is seldom found with all the branches complete; hydrothecae alternate, stout, narrowing gradually but slightly, towards the circular opening, with no very distinct neck; the margin, which is even, lies parallel to the axis of the stem or branch; hydrothecae well immersed, seldom more than one-fourth being free.

Gonosome.—Gonangia are borne on the upper surface of the branches; elongated oval, with a distinct pedicel; collar short but distinct; aperture large; surface smooth or very slightly wrinkled.

Distribution.—Entrance to strait of Juan de Fuca (Nutting); San Juan archipelago; Nawhitti bar, Departure bay, off Matia island, off Waldron island, off O'Neale island, Friday Harbor, San Juan channel, Upright channel, Deer harbour, Griffin bay; Houston Stewart channel in the western portion and off Rose harbour (Fraser); Queen Charlotte sound, Nawhitti bar, western portion of Goose island halibut grounds, off Sadler point, off Klashwan point, off Massett inlet, 5 miles east of Sand spit; off McArthur reef and off Shingle island in Sumner strait, Alaska. 8 to 240 fathoms.

Abietinaria traski (Torrey)
Plate 29, Fig. 156

Sertularia traski TORREY, Hyd. of the Pacific coast, 1902, p. 69.
Abietinaria traski NUTTING, Am. Hyd., II, 1904, p. 118.
FRASER, West Coast Hyd., 1911, p. 63.
Vancouver Island Hyd., 1914, p. 182.
Hyd. from west coast of V.I., 1935, p. 145.

Trophosome.—Stem long, straight, even in the younger colonies the proximal half is free of branches, and, in the older ones, the branched portion is relatively small as compared with the unbranched portion; branches very regularly pinnate and graded so well in length that the colony is very symmetrical and graceful; although the stem is a light horn colour, the branches are quite white. The nodes are not regular in the main stem and are absent, or nearly so, in the branches; hydrothecae alternate, rather distant, short and thick, with a short neck; margin not expanded, nearly circular but straight on the adcauline side.

Gonosome.—Gonangia arranged in rows on the face of the branches; oblong, without a distinct pedicel or collar; surface smooth.

Distribution.—San Pedro (Torrey); entrance to strait of Fuca, Oregon coast, off Monterey, Cal. (Nutting); San Juan archipelago, Dodds narrows, Departure bay; China Hat, Clayoquot sound, Swiftsure shoal, off Lasqueti island, Nanoose bay, off Clarke rock, Departure bay, north of Gabriola island, Northumberland channel, Dodds narrows, Gabriola reefs, Whaleboat passage, off Matia island, off Waldron island, off O'Neale island, off Brown island, Upright channel, Griffin bay, Deer harbour, Port Townshend, Puget sound; off Clayoquot sound, off Sydney inlet (Fraser); San Pedro, locations in each of the three sections of San Francisco bay; west of Nawhitti bar, off Virago sound, off Massett inlet, off cape James, Hope island; Queen Charlotte sound; Berg inlet, off Easterly island in Ernest sound, off McArthur reef in Sumner strait, Alaska. 5 to 204 fathoms.

Abietinaria turgida (Clark)
Plate 30, Fig. 157

Thuiaria turgida CLARK, Alaskan Hyd., 1876, p. 229.
Abietinaria turgida NUTTING, Am. Hyd., II, 1904, p. 115.
FRASER, West Coast Hyd., 1911, p. 63.
Vancouver Island Hyd., 1914, p. 115.
Hyd. Distr. in vicinity of Q.C.I., 1936, p. 125.

Trophosome.—Stem straight, stout, distal portion branched; branches irregularly pinnate, constricted at the base, similar to the main stem; branched again, either alternately or dichotomously; internodes on main stem short, on branches, long; hydrothecae nearly opposite, crowded, tubular, with a very slight narrowing distally, almost wholly immersed; margin entire, circular, at right angles to the hydrotheca.

Gonosome.—Gonangia crowded on the stem and proximal portions of the branches; large, oval, with a short collar; aperture small, pedicel short; distinct longitudinal crests present.

Distribution.—Alaskan coast, Aleutian islands and Bering sea (Nutting); Orca, Alaska; Gabriola pass, off Matia island, Port Townshend; Danger rocks at the eastern entrance and in the western portion of Houston Stewart channel; entrance to Flamingo harbour, entrance to Big bay, outside Massett inlet (Fraser); in each section of San Francisco bay; off Heceta head, Ore., Jamestown bay, Whale island; off cape James, Hope island; Sitka, Orca, Kadiak, Alaska. Low tide to 76 fathoms.

Abietinaria variabilis (Clark)
Plate 30, Fig. 158

Sertularia variabilis CLARK, Alaskan Hyd., 1876, p. 221.
Abietinaria variabilis NUTTING, Am. Hyd., II, 1904, p. 123.
FRASER, West Coast Hyd., 1911, p. 63.
Vancouver Island Hyd., 1914, p. 183.
Hyd. from west coast of V.I., 1935, p. 145.
Hyd. Distr. in vicinity of Q.C.I., 1936, p. 125.

Trophosome.—Stem stout, straight or slightly flexuous; branches regularly alternate, stout and long, constricted at the base; hydrothecae alternate, much the same size for the proximal two-thirds or three-fourths, which is immersed, and then suddenly narrowing to form a neck; the margin very little flaring, oblique relative to the stem, but nearly horizontal.

Gonosome.—Gonangia borne on the upper sides of the branches; obovate, without collar; aperture large; surface smooth.

Distribution.—San Miguel island, off Oregon coast, off entrance to strait of Fuca, east and west of Kodiak island, Aleutian islands, Bering sea (Nutting); Bering sea (Jäderholm); Queen Charlotte islands; off cape Edenshaw, off Massett, off Rose spit; Swiftsure shoal, Friday Harbor; off Long beach, off Clayoquot sound; western portion of Houston Stewart channel (Fraser); North head, Ore.; west of Nawhitti bar, off Tlell; Kadiak, Alaska. 8 to 171 fathoms.

Genus DICTYOCLADIUM

Trophosome.—"Colony flabellate in form. Branches anastomosing and forming a widely reticulated structure or network. Hydrothecae on more than two sides of the stem. Aperture without conspicuous teeth, operculum variable" (Nutting).

Gonosome.—"Gonangia borne in the bifurcations of the branches and marked with annular rugosities" (Nutting).

Dictyocladium flabellum Nutting
Plate 30, Fig. 159

Dictyocladium flabellum NUTTING, Am. Hyd., II, 1904, p. 105.

FRASER, West Coast Hyd., 1911, p. 63.

Trophosome.—"Colony flabellate in form, attaining a height of about 4 inches and branching in a strictly dichotomous manner; few evident internodes on stem or branches; the only annulations or constrictions ordinarily being those at the origin of branches or branchlets. Branches straight, not flexuous, themselves dichotomously branching in the same plane, the ultimate branches often anastomosing with other branches, forming a rude reticulate pattern. Hydrothecae arranged in four longitudinal series on stem and branches, so as to form an ascending spiral; tubular, about the distal one-third free, curved gently outward, margin irregular, but usually with quadrilateral outline, with the corners of the outline very slightly if at all produced into four very low, obscure teeth; operculum with four flaps" (Nutting).

Gonosome.—"Gonangia borne in bifurcations of the branches, very large, ovate, body with shallow, broad, obscure annulations; neck in the form of a low truncated cone, with a round, terminal aperture" (Nutting).

Distribution.—Off entrance to strait of Fuca, north of Unimak pass, 27-72 fathoms (Nutting).

Genus DIPHASIA

Trophosome.—Hydrothecae in two rows on the stem and branches; operculum of a single adcauline flap.

Gonosome.—Gonangia commonly marked with spines or lobes; an internal marsupium is usually present in the female.

KEY TO SPECIES

A. Branches arising from all sides of the stem........*D. pulchra*
B. Branching pinnate
    a. Gonangia with two or more lateral spines...*D. corniculata*
    b. Gonangia without spines..................*D. kincaidi*

Diphasia corniculata (Murray)
Plate 30, Fig. 160

Sertularia corniculata MURRAY, Ann. and Mag. N.H., (3), V, 1860, p. 251.

Diphasia corniculata NUTTING, Am. Hyd., II, 1904, p. 112.

FRASER, West Coast Hyd., 1911, p. 64.

Trophosome.—"Cells not quite opposite, sometimes nearly alternate, forming an open cup, resting on the stem; lip not distinct; exterior margin somewhat projecting at tip; a single one in the axile of each pinna" (Nutting, after Murray).

Gonosome.—"Vesicles pearshaped, with two long points projecting like horns, at the thick end, aperture between them" (Nutting, after Murray).

Distribution.—Bay of San Francisco (Murray).

Diphasia kincaidi (Nutting)
Plate 30, Fig. 161

Thuiaria elegans NUTTING, Hyd. of the Harriman Exp., 1901, p.187.
Thuiaria kincaidi NUTTING, Am. Nat., 1901, p. 789.
Diphasia kincaidi NUTTING, Am. Hyd., II, 1904, p. 112.
FRASER, West Coast Hyd., 1911, p. 64.
Hyd. of Miramichi, 1926, p. 209.

Trophosome.—Stem stout, irregularly branched; branches with no secondary branches in the proximal portion but with numerous branches distally, giving a plumose effect to the colony; stem divided into irregular internodes by oblique nodes; internodes of branches also irregular and nodes oblique; hydrothecae on stem and branches alternate or subalternate; the base of each hydrotheca some distance above the margin of the preceding hydrotheca on the same side; short, stout, pitcher-shaped, both as to the whole hydrotheca and the margin; operculum somewhat vaulted to fit the sinuous margin.

Gonosome.—Gonangia in double rows along the distal portions of the stem and branches; small for the genus, oblong-ovate, sessile; distal and truncate; aperture rounded; no spines or external projections.

Distribution.—Berg inlet and Dutch harbour, Alaska (Nutting); Berg inlet, Point Gardiner buoy off Admiralty island, Alaska. 29 fathoms.

Diphasia pulchra Nutting
Plate 30, Fig. 162

Diphasia pulchra NUTTING, Am. Hyd., II, 1904, p. 111.
FRASER, West Coast Hyd., 1911, p. 64.

Trophosome.—Stem slender, somewhat geniculate distally; branches coming off from all sides of the stem, spirally arranged, forming a dense bushy tuft, not unlike that of *Thuiaria argentea*;

hydrothecae alternate, rather distant, long pitcher-shaped, slender; margin with two broad teeth, opposite to each other.

Gonosome.—Undescribed. As it has an internal marsupium, the species definitely belongs to the genus *Diphasia.*

Distribution.—Off the entrance to the strait of Fuca, 67 fathoms (Nutting).

Genus HYDRALLMANIA

Trophosome.—Hydrothecae in groups on the sides of the stem or branches, their bases in line but their distal ends turned alternately right and left. Operculum of a single adcauline flap.

Gonosome.—Gonangia without spines or internal marsupium.

KEY TO SPECIES

A. Hydrotheca distinctly flask-shaped, distal end much constricted, aperture circular...............*H. franciscana*

B. Hydrotheca more nearly tubular, distal end not distinctly constricted, aperture not circular...............*H. distans*

Hydrallmania distans Nutting

Plate 31, Fig. 163

Hydrallmania distans NUTTING, Hyd. from Alaska and Puget sound, 1899, p. 746.

NUTTING, Am. Hyd., II, 1904, p. 126.

FRASER, West Coast Hyd., 1911, p. 65.

Vancouver Island Hyd., 1914, p.185.

Hyd. of west coast of V.I., 1935, p. 145.

Hyd. Distr. in vicinity of Q.C.I., 1936, p. 125.

Trophosome.—Stem erect, with a large proximal portion free of branches; branching much varied, often distinctly bilateral but at times with branches coming off in all planes. When the colony is bilateral, the branches are regularly pinnate and unbranched, making the colony look trim, but when the branches come off in all planes, these branches are often branched alternately or dichotomously to give a very shaggy appearance. Hydrothecae inserted with their bases in line and their distal ends alternately turned to right and left but this feature is not so noticeable in the younger specimens; the alignment may be so much out that it does not give the characteristic appearance of the adult colony. The length of the internodes is not definite, the number of hydrothecae on each varying from three upwards; hydrothecae tubular with the distal end pitcher-shaped like that of a *Diphasia* and, like it also, with operculum of one adcauline flap.

Gonosome.—Gonangia borne on front of branches, sometimes forming a definite row, oval or obovate, with short neck, wide aperture and distinct pedicel; surface smooth.

Distribution.—Puget sound (Calkins); Puget sound (Nutting); almost everywhere in the Vancouver island and Queen Charlotte islands regions, where dredging has been done (Fraser); locations in the middle and lower sections of San Francisco bay; several further records in the Queen Charlotte islands region. 3 to 75 fathoms.

Hydrallmania franciscana (Trask)
Plate 31, Fig. 164

Plumularia franciscana TRASK, Proc. Cal. Acad. Sc., 1857, p. 113.
Hydrallmania franciscana NUTTING, Am. Hyd., II, 1904, p. 126.
FRASER, West Coast Hyd., 1911, p. 65.

Trophosome.—"Polypidom six or eight inches high, color corneous; alternately branched, the branches pinnated, one branch to each internode of the stem. The pinnae arise one above the other, are pointed and support three cells at each joint. On two specimens, four cells have been met with, but this may be regarded as an exception rather than otherwise. The pinnae are dichotomously branched in adult specimens. Cells lageniculate, smooth, free, slightly decumbent, the attachment of the base is marked by a slightly elevated, rounded rim; apertures round and smooth" (Trask).

Distribution.—San Francisco bay (Trask and Murray).

Hydrallmania distans is plentiful in the San Francisco bay region but no *Hydrallmania* appeared in the extensive San Francisco bay material that answers to the description of *H. franciscana* as given by Trask.

Genus SELAGINOPSIS

Trophosome.—Hydrothecae arranged in more than two longitudinal series, at least on distal part of the branches, or in two or more series each of which has the distal ends of the hydrothecae turned alternately to right and left. Operculum usually of a single abcauline flap.

Gonosome.—Gonangia oval or obovate, usually smooth or nearly so.

KEY TO SPECIES

A. Distal ends of the hydrothecae in the one series turning alternately to the right and left
   a. Hydrothecae in two series...............*S. alternitheca*
   b. Hydrothecae in four series.................*S. hartlaubi*
B. Distal end of the hydrothecae in any series in alignment with the base
   a. Hydrothecae in three longitudinal series
      a. Hydrotheca almost wholly immersed.....*S. triserialis*
      b. Hydrotheca about one-fourth free.......*S. trilateralis*
   b. Hydrothecae in four longitudinal series
      a. Primary branches unbranched; stem stiff and woody..
        ....*S. pinnata*
      b. Primary branches often branched
        1. A network of canals in the stem and four canals in the branches; gonangia with long curved processes...........................*S. ornata*
        2. No distinct coenosarcal canals; gonangia transversely annulated................*S. cedrina*
   c. Hydrothecae in six or more longitudinal series
      a. Distal ends of hydrothecae distinctly exserted
        1. Hydrothecal margin with two distinct teeth......
        ....*S. mirabilis*
        2. Hydrothecal margin without teeth.....*S. pinaster*
      b. Distal ends of hydrothecae not distinctly exserted
        1. Hydrothecae in four rows proximally and six rows distally......................*S. cylindrica*
        2. Hydrothecae in six or eight rows.......*S. obsoleta*

Selaginopsis alternitheca (Levinsen)

Plate 31, Fig. 165

Thuiaria alternitheca LEVINSEN, Vid. Meddel. naturh. Foren., 1892, p. 32.
Selaginopsis alternitheca NUTTING, Am. Hyd., II, 1904, p. 133.

Trophosome.—(Description from a fragment 2.5 cm. long). Stem and branches rigid; nodes distant, irregularly placed; branches given off alternately, three hydrothecae on the stem between two successive branches on the same side. Each branch arises from a distinct process of the stem, the joint being definite; no definite nodes on the branch. There is a double row of hydrothecae on each side of the branch, each row with the bases of the hydrothecae in line, but the distal portions turned alternately to right and left. Hydrothecae of nearly the same diameter throughout, most of the curvature being in the distal half; margin entire; operculum of one abcauline flap.

Gonosome.—"Gonangia borne on basal portions of branches, elongated oval, abruptly truncated at distal end, with a very broad aperture and no neck" (Nutting).

Distribution.—Off Frederick island, Q.C.I. 15 fathoms.

Selaginopsis cedrina (Linnaeus)
Plate 31, Fig. 166

Sertularia cedrina LINNAEUS, Syst. Nat., 1758, p. 814.
Selaginopsis pacifica MERESCHKOWSKY, Ann. and Mag. N.H., (5), II, 1878, p. 438.
Selaginopsis cedrina NUTTING, Am. Hyd., II, 1904, p. 130.
FRASER, West Coast Hyd., 1911, p. 65.

Trophosome.—"Hydrocaulus slightly curved, divided into regular internodes. Branches arranged alternately on two sides of the principal stem, two pairs on each internode, divided into five internodes, constricted at the point of attachment and at the nodes. Each branch bears one or two, rarely five, secondary branches. Hydrothecae cylindrical, almost entirely immersed in the substance of the axial tube; aperture oval, with two angles (not teeth); hydrothecae arranged in four regular series, and at the same time in a spiral, the hydrothecae of each series following one another immediately, without leaving any free space or interval" (Mereschkowsky).

Gonosome.—"Gonangia arranged in two or three series, of an oval form, narrowing gradually towards the base, and truncate at the apex. The surface is ribbed" (Mereschkowsky).

Distribution.—Bering sea (Kirchenpauer).

Selaginopsis cylindrica (Clark)
Plate 31, Fig. 167

Thuiaria cylindrica CLARK, Alaskan Hyd., 1876, p. 226.
Selaginopsis cylindrica NUTTING, Am. Hyd., II, 1904, p. 131.
FRASER, West Coast Hyd., 1911, p. 65.
Vancouver Island Hyd., 1914, p. 187.
Hyd. Distr. in vicinity of Q.C.I., 1936, p. 125.

Trophosome.—Stem rather stout but more so distally than proximally; this is also true of the branches; branches regularly alternate and graded in such a way as to give a plumose appearance to the colony; sometimes the branches are curved or spirally twisted and have secondary branches to such extent as to make the colony look bushy.   In older colonies, the main stem often becomes quite woody and some of the primary branches become so large that they look like main stems, with pinnate branching in the same way.   Hydrothecae tubular, almost wholly immersed, arranged in two series on the main stem, 4 series on the proximal portions of the branches and 6 series on the distal portions, the two extra series being intercalated at the same time, often quite near the base; aperture oval; margin entire; operculum of one abcauline flap.

Gonosome.—(Not previously described).   Gonangia appear in a closely continuous row, or there may be two rows, on the upper side of the branches; rather large, obliquely pear-shaped, but the margin is at right angles to the axis, opening occupying the whole distal end; usually a slight indication of a collar.

Although this is a common species for the northern part of the coast, gonangia were observed for the first time on a colony, 11 cm. high, dredged in Burrard inlet, B.C., in 15 fathoms, Dec. 21, 1932.

Distribution.—Port Moller, Hagmeister island, Bering sea, Chirikoff islands, Chiachi islands, Alaska (Clark); Puget sound (Calkins); Bristol bay, Alaska (Nutting); St. Lawrence island, Bering sea (Jäderholm); San Juan archipelago, Queen Charlotte islands; off cape Edenshaw, off Massett, Rose spit, Departure bay, off Matia island, off O'Neale island, San Juan channel, Friday Harbor, Griffin bay, Deer harbour; Houston Stewart channel in the western portion and off Rose harbour, Massett harbour, outside Massett inlet (Fraser); Port Orchard, Puget sound; off Massett inlet, east of Rose spit, off Tlell, 5 miles east of Sand spit; 16 miles northeast of Reef island, off cape James, Hope island.   8 to 50 fathoms.

Selaginopsis hartlaubi Nutting

Plate 31, Fig. 168

Selaginopsis hartlaubi NUTTING, Am. Hyd., II, 1904, p. 133.
                    FRASER, West Coast Hyd., 1911, p. 66.
                    Vancouver Island Hyd., 1914, p. 187.

Trophosome.—Stem stout, irregularly branched, branches varying very much in length, almost as stout as the main stem, not branched again; hydrothecae in two regular series on the stem but in four series on the branches; in each series on the branches, the distal ends of the hydrothecae turn to right and left alternately and the bases are seldom perfectly in line; hydrothecae immersed, tubular, with distal ends constricted; margin entire, oval; operculum of one abcauline flap.

Gonosome.—Unknown.

Distribution.—South of Pribilof islands, 43 fathoms (Nutting); San Juan archipelago; off O'Neale island (Fraser).

Selaginopsis mirabilis (Verrill)
Plate 32, Fig. 169

Diphasia mirabilis VERRILL, Amer. Jour. Sc., (3), 5, 1872, p. 9.
Selaginopsis mirabilis NUTTING, Am. Hyd., II, 1904, p. 128.
  FRASER, West Coast Hyd., 1911, p. 66.
    Vancouver Island Hyd., 1914, p.188.
    Hyd. from west coast of V.I., 1935,
      p. 145.
    Hyd. Distr. in vicinity of Q.C.I.,
      1936, p. 125.

Trophosome.—Stem stout, with regular pinnate branching, slightly geniculate in the branched portion; hydrothecae in two rows on the stem but in six rows on the branches; distal portion of the hydrotheca free and turned outward from the stem; margin oval, with two lateral teeth; operculum of two flaps.

Gonosome.—Gonangia oval, not constricted to form a distinct neck; aperture large, circular; surface smooth; a distinct pedicel is present.

Distribution.—Hagmeister island, Bering sea, Popoff strait, Shumagin islands, Alaska (Clark); Puget sound, entrance to the strait of Fuca (Nutting); San Juan archipelago, Dodds narrows; off Massett, Northumberland channel, Pylades channel, Gabriola pass, Ruxton passage, off Matia island, Friday Harbor, off Brown island, Deer harbour, Upright channel, Griffin bay, off Waldron island, off O'Neale island, off Blakeley island, Puget sound; off Clayoquot sound, off Bajo reef; outside Massett inlet (Fraser); off Klashwan point, off Massett inlet, off Tlell, Q.C.I. 7 to 45 fathoms.

10

Selaginopsis obsoleta (Lepechin)

Plate 32, Fig. 170

Sertularia obsoleta LEPECHIN, Acta Acad. Petropol., (2), II, 1778, p. 157.

Selaginopsis obsoleta NUTTING, Am. Hyd., II, 1904, p. 132.

FRASER, West Coast Hyd., 1911, p. 66.

Trophosome.—"Colony attaining a height of about 4 inches. Stem thick, greatly geniculate, divided into irregular internodes, the tendency being towards an arrangement in which there are two branches to an internode, with an occasional deeply cut node, particularly on the distal portion, and also very shallow annulations that are much more numerous than the real nodes. Branches alternate, closely approximated, springing from short processes on the stem, from which they are separated by very deep nodes; otherwise the nodes are almost entirely absent. Hydrothecae arranged in six regular series, so that they form both vertical rows and spirals, tubular, rather short, broader at the base and narrowing distally to the smooth margin and oval aperture; there are no marginal teeth, and the operculum consists of a single abcauline flap. There is usually a distinct space intervening between the top of one hydrotheca and the bottom of the one immediately above it" (Nutting).

Gonosome.—Gonophores in the form of a reversed cone, larger and more elongated in the adult than in the young stage. A short pedicel present; distal and truncate with the margin rounded, furnished with a tube so short as to be almost unobservable. (Condensed from Mereschkowsky.)

Distribution.—St. Paul's island, Bering sea (A. and A. Krause); south of Nunivak island, Bering sea, 23 fathoms (Nutting).

Selaginopsis ornata Nutting

Plate 32, Fig. 171

Selaginopsis ornata NUTTING, Am. Hyd., II, 1904, p. 131.

FRASER, West Coast Hyd., 1911, p. 66.

Trophosome.—Colony reaching a height of 10 cm. Stem straight, stout, rigid, with numerous stout branches pinnately arranged, but not always regularly so. Distinct nodes distant and irregularly spaced; coenosarc canaliculated; branches borne on short but distinct processes of the stem, with two or three hydrothecae between two successive branches on the same side of the stem; commonly unbranched but occasionally with branchlets in the distal portion;

coenosarc of the branches also canaliculated; hydrothecae in four regular, equidistant rows; almost regularly tubular, but curved so that the free margin is turned outward; almost wholly immersed; aperture oval, the lateral teeth quite shallow; operculum of a single abcauline flap.

Gonosome.—"Gonangia borne in rows on front of branches, each being inserted just below the base of a hydrotheca; obconical, long, protruded in a rather slender pedicel below, and having about eight remarkably long, bifurcated arms or processes above, which curve inward towards each other at their distal ends so as to form a sort of pseudo-marsupium, above the body of the gonangium" (Nutting).

Distribution.—Off Chignik bay, Alaska, 45 fathoms (Nutting); off Sadler point, off Massett inlet, Q.C.I.   30-45 fathoms.

Selaginopsis pinaster (Lepechin)
Plate 32, Fig. 172

Sertularia pinaster LEPECHIN, Acta Acad. Petropol., 1783, p. 223.
Sertularia pinus KIRCHENPAUER, Nordische Gattungen, 1884, p. 11.
Selaginopsis pinaster NUTTING, Am. Hyd., II, 1904, p. 128.
FRASER, West Coast Hyd., 1911, p. 66.

Trophosome.—Stem usually simple but sometimes slightly fascicled towards the base; cylindrical, arising from a creeping network to the height of 6 inches; dark brown at the base but becoming lighter distally.   Branching irregular, the branches slender and flexible, often supporting six rows of hydrothecae.   Hydrothecae oval but narrowing distally to form a slight neck, terminated by an opening.   (Adapted from Kirchenpauer.)

Gonosome.—Gonangia tubular, generally tumid, almost transparent, with a circular opening, surrounded by a thickened border; closely crowded, almost imbricate, on both sides of the branches. (After Kirchenpauer.)

Distribution.—St. Paul's island, Bering sea (A. and A. Krause).

Selaginopsis pinnata Mereschkowsky
Plate 32, Fig. 173

Selaginopsis pinnata MERESCHKOWSKY, Ann. and Mag. N.H., (5),
II, 1878, p. 435.
NUTTING, Am. Hyd., II, 1904, p. 130.
FRASER, West Coast Hyd., 1911, p. 66.
Vancouver Island Hyd., 1914, p. 188.

Trophosome.—Stem stout, woody, with deeply cut nodes; branches not regularly alternate; they vary from alternate to opposite; hydrothecae in two rows on the stem and four very regular rows on the branches, not so closely crowded as in some of the other species; tubular, but short and stout, with oval aperture; operculum with one abcauline flap.

Gonosome.—Unknown.

Distribution.—Port Ajan (M. Wosnessensky); St. Paul's island, 23 fathoms (Kirchenpauer); south of Pribilof islands, 25 fathoms (Nutting); San Juan archipelago, Queen Charlotte islands; off Rose spit, Alert bay (Fraser).

Selaginopsis trilateralis Fraser
Plate 33, Fig. 174

Selaginopsis trilateralis FRASER, Queen Charlotte Hyd., 1936, p.504.

Trophosome.—A rigid colony, 18 mm. high, is regularly, pinnately branched, the branches constricted proximally; hydrothecae in two series on the greater portion of the stem but in three distinct series on the branches; hydrothecae about one-fourth free; margin with two, low, rounded teeth; operculum of two flaps.

Gonosome.—Unknown.

Distribution.—In Houston Stewart channel, near Rose harbour, 30 fathoms, and in the western portion of the channel, 15-18 fathoms (Fraser).

Selaginopsis triserialis Mereschkowsky
Plate 33, Fig. 175

Selaginopsis triserialis MERESCHKOWSKY, Ann. and Mag. N.H.,(5), II, 1878, p. 435.

NUTTING, Am. Hyd., II, 1904, p. 129.

FRASER, West Coast Hyd., 1911, p. 66.

Vancouver Island Hyd., 1914, p.188.

Hyd. Distr. in vicinity of Q.C.I., 1936, p. 125.

Trophosome.—Stem rather slender for the genus, reaching a height of 30 cm., with long branches arranged alternately but not regularly; hydrothecae in two rows on the stem and sometimes also on the proximal portions of the branches but in three rows in the distal portions; more distant than in other species of the genus; almost wholly immersed; aperture oval; operculum a single abcauline flap.

Gonosome.—"Gonangium oblong-oval in shape, with a large terminal aperture" (Nutting).

Distribution.—San Pedro (Torrey); off point Conception, Cal., 31 fathoms (Nutting); Nanoose bay, off Waldron island (Fraser); off San Bruno light and off Southampton light, in San Francisco bay. 1¼ to 31 fathoms.

Genus **SERTULARELLA**

Trophosome.—Hydrothecae in two rows, alternate, usually with three or four teeth and an operculum of three or four flaps.

Gonosome.—Gonangia usually supplied with ridges or corrugations.

KEY TO SPECIES

A. Hydrothecal margin with four teeth
   a. Hydrothecae wholly immersed.................*S. albida*
   b. Hydrothecae partly free
     a. Hydrothecal walls annulated
       1. Annulations distinct on adcauline wall only......
        ....*S. conica*
       2. Annulations complete
         i. Hydrothecae decidedly rugose......*S. rugosa*
         ii. Hydrothecae less rugose
           I. Stem slender, geniculate.......*S. tenella*
           II. Stem stout, nearly straight....*S. tanneri*
     b. Hydrothecal walls smooth
       1. About 2/3 of adcauline wall adnate...*S. complexa*
       2. About 1/2 of adcauline wall adnate..*S. polyzonias*
       3. Less than 1/3 of adcauline wall adnate.........
        ....*S. fusiformis*
B. Hydrothecal margin with three teeth
   a. Hydrothecal wall annulated...............*S. pedrensis*
   b. Hydrothecal wall smooth
     a. Teeth unequal
       1. Stem erect, stout...................*S. turgida*
       2. Stem flabellate....................*S. pinnata*
     b. Teeth equal
       1. Distal portion of hydrotheca expanding..*S. elegans*
       2. Distal and proximal portions of hydrotheca nearly equal
         i. Gonangia with prominent compressed annular ridges
           I. Colony fragile and delicate....*S. minuta*

II. Colony lax but more robust...........
...................*S. tricuspidata*
ii. Gonangia with shallow annulations...*S. levinseni*
C. Hydrothecal margin with two teeth...............*S. clarki*
D. Marginal teeth very inconspicuous; hydrotheca large.*S. magna*

Sertularella albida Kirchenpauer
Plate 33, Fig. 176

Sertularella albida KIRCHENPAUER, Nordische Gattungen, 1884, p. 42.
NUTTING, Am. Hyd., II, 1904, p. 86.
FRASER, West Coast Hyd., 1911, p. 68.
Vancouver Island Hyd., 1914, p. 190.
Hyd. Distr. in vicinity of Q.C.I., 1936, p. 125.

Trophosome.—Stem stout, branching irregular but tending to be alternate; branches often branched again and as they are not uniform, the whole colony has a ragged, spreading appearance; nodes in both stem and branches deeply cut but irregularly placed; hydrothecae tubular, curved slightly on the adcauline side, and having a distinct bend on the abcauline side, placed closely together and wholly immersed, or nearly so; margin with four teeth that are not very conspicuous; operculum with four flaps.

Gonosome.—"Gonangia axillary, very large, perhaps the largest found in the genus, ovate, regularly and closely annulated, with short, tubular neck and round terminal aperture" (Nutting).

Distribution.—Yukon harbour, Big Koniushi, Shumagin islands, Alaska, 6-20 fathoms (Clark); off Matia island, off Waldron island; western portion of Houston Stewart channel, 15-18 fathoms(Fraser).

Sertularella clarki Mereschkowsky
Plate 33, Fig. 177

Sertularella clarki MERESCHKOWSKY, Ann. and Mag. N.H., (5), II, 1878, p. 447.
NUTTING, Am. Hyd., II, 1904, p. 102.
FRASER, West Coast Hyd., 1911, p. 68.

Trophosome.—"Hydrorhiza forming a compact layer of hydrophytons. Hydrocaulus straight, long, cylindrical, not angularly bent, with regular internodes, destitute of branches to the apex, where the width of the axial tube suddenly diminished considerably, and it, at the same time, gives origin to branches. Branches divided into internodes, rather short, issuing from all sides of the principal stem, one from each of the internodes, ramified in their turn so that each branch internode gives off a secondary branch, which is divided once or twice; and all these secondary branches are turned towards the axis of the colony (inwards). Hydrothecae tubular, a little contracted at the extremity; aperture broad, oval, furnished with two large teeth arranged unsymmetrically; arrangement of the hydrothecae, although biserial, not in the same plane, having, at the first glance, the appearance of being uniserial" (Mereschkowsky).

Gonosome.—Unknown.

Distribution.—Unalaska (M. Pètelin).

Sertularella complexa Nutting
Plate 33, Fig. 178

Sertularella complexa NUTTING, Am. Hyd., II, 1904, p. 94.

FRASER, West Coast Hyd., 1911, p. 68.

Trophosome.—Colony straggling, with little difference between stem and branches; both long and slender, slightly flexuous at the nodes that are usually distinct, placed obliquely; the internodes being short and bearing but one hydrotheca. The branches make a wide angle with the stem; at times the terminal branches are dichotomous. A certain amount of anastomosis may take place between the stem and branches or between the branches; hydrothecae distant, approximately one-third free; the basal portion tapers towards the abcauline side to make an acute angle; margin with four low teeth; operculum with four flaps.

Gonosome.—Gonangia borne in rows on the face of the stem and branches; oval or ovate, with broad, rounded annulations; aperture round; margin surrounded by an elevation, supporting three to seven (usually 4 or 5) blunt teeth.

Distribution.—Off Prince William sound, 230 fathoms, south of Kodiak island, 159 fathoms, south of Unimak pass, 45 fathoms, south of Pribilof islands, 121 fathoms, all in Alaskan waters (Nutting).

Sertularella conica Allman
Plate 34, Fig. 179

Sertularella conica ALLMAN, Hyd. of the Gulf Stream, 1877, p. 21.

NUTTING, Am. Hyd., II, 1904, p. 79.

FRASER, West Coast Hyd., 1911, p. 68.

Vancouver Island Hyd., 1914, p. 190.

Sertularella conella STECHOW, Zool. Anzeiger, 1921, p. 231.
Sertularella conica FRASER, Hyd. from west coast of V.I., 1935,
     p. 145.
          Hyd. Distr. in vicinity of Q.C.I., 1936,
               p. 126.

Trophosome.—Colony small, either unbranched or with a few small branches, which are like the main stem; hydrothecae alternate, rather distant, free for about two-thirds of their length, nearly tubular, but with the proximal end slightly swollen and the distal end narrowing to some extent; an appearance of annulations but these only on the adcauline side of the hydrothecae; margin with four teeth; operculum with four flaps.

Gonosome.—Gonangia on the stem or on the stolon; oval, without distinct pedicel or neck; margin provided with three or four stout teeth, that may be straight or curved inward almost to meet above the centre of the aperture; surface rugose, with distinct crest on the rugosities.

There is much variation in the appearance of the gonangia, especially in the teeth; in specimens from the Atlantic ocean and all the way up the Pacific coast of Central America, Mexico, and farther north, all gradations between the extremes appear. There is, therefore, no necessity of creating a new species for the west coast forms in the area under consideration, as Stechow suggests.

Distribution.—Townshend harbour (Calkins); San Juan archipelago, Port Renfrew, Ucluelet; Northumberland channel, Dodds narrows, Gabriola pass, Friday Harbor, Swiftsure shoal, Claninnick; Bajo reef, Nootka island; east shore of Flamingo harbour, Danger rocks at the eastern entrance, and in the western portion of Houston Stewart channel (Fraser); San Diego, Cal.; Heceta head, Ore.; Dewey anchorage at Etolin island, Alaska. Low tide to 67 fathoms.

Sertularella elegans Nutting
Plate 34, Fig. 180

Sertularella elegans NUTTING, Am. Hyd., II, 1904, p. 98.
          FRASER, West Coast Hyd., 1911, p. 69.

Trophosome.—"Colony growing from a rootstock parasitic on *Abietinaria,* and attaining a height of about 3 inches. Stem not fascicled, with several strong annulations on proximal portion, divided into regular internodes, each bearing a hydrotheca which is directed forward, outward and upward; nodes very strong and deeply cut. Branches straggling and irregular, tending to an alternate arrangement, and sometimes anastomosing, divided into deeply incised nodes into rather short, regular internodes, each of which bears a hydrotheca. Hydrothecae directed upward, forward and outward, rather closely approximated, tubular, gently curved, adherent by about their proximal, adcauline third; margin expanding, with three strong and equidistant teeth, and with a narrow border or rim; operculum of three flaps. Hydrothecae often with an oblique intrathecal ridge running downward from the anterior margin" (Nutting).

Gonosome.—"Gonangia in rows on stems and main branches, exceedingly elaborate in ornamentation, oval to round, neck tubular with trumpet-shaped aperture, the usual annulations produced into raised, fluted frills that look like a series of lace collars, giving an exceedingly ornate effect" (Nutting).

Distribution.—South of Unimak pass, Alaska, 72 fathoms (Nutting).

Sertularella fusiformis (Hincks)

Plate 34, Fig. 181

Sertularia fusiformis HINCKS, Ann. and Mag. N.H., (3), VIII, 1861, p. 253.

Sertularella fusiformis HINCKS, Br. Hyd. Zooph., 1868, p. 243.

NUTTING, Am. Hyd., II, 1904, p. 89.

FRASER, West Coast Hyd., 1911, p. 69.

Trophosome.—Stem slender, less than 25 mm. in height, slightly flexuose, unbranched or slightly branched; the branch resembling the main stem, divided into regular nodes, each with one hydrotheca.   Besides the node there may be one or more annulations at the base of the internode; the axis of the stem is somewhat twisted at each node. Hydrothecae rather distant, elongate, somewhat urceolate, but slightly adherent to the stem; opening nearly square with a four-toothed margin; operculum with four flaps.

Gonosome.—Gonangia few in number, sometimes only one or two, irregularly placed; elongated oval or elliptical, with definitely crested, transverse rugosities.   There is a short tubular neck bearing the circular aperture, the margin of which is provided with four blunt teeth.

Distribution.—San Francisco (Torrey); off San Bruno light, Alcatraz island and Southampton light, in San Francisco bay; three locations off Heceta head, Ore. 6 to 62 fathoms.

Sertularella levinseni Nutting
Plate 34, Fig. 182

Sertularella levinseni NUTTING, Am. Hyd., II, 1904, p. 100.

FRASER, West Coast Hyd., 1911, p. 69.

Trophosome.—"Colony very loose and straggling in habit, sometimes attaining a height of 3 inches. Stem not fascicled, slender, flexuose, divided into regular internodes, each of which bears a hydrotheca or a hydrotheca and a branch. Branches regularly alternate, slender, flexuose, often dividing dichotomously, rarely anasomosing, divided into regular internodes throughout. Hydrothecae rather small and distant, stout, swollen below, their adcauline wall adnate for from one-half to three-fourths of its length; margin with three well-marked equal and equidistant teeth; operculum with three flaps" (Nutting).

Gonosome.—"Gonangia borne in rows on stem and all the branches, although they are more apt to be aggregated proximally; small, ovoid, somewhat elongate, with shallow, broad annulations, particularly on distal portions; neck short but distinct" (Nutting).

Distribution.—South of Unimak pass, Alaska, 72 fathoms (Nutting).

Sertularella magna Nutting
Plate 35, Fig. 183

Sertularella magna NUTTING, Am. Hyd., II, 1904, p. 103.

FRASER, West Coast Hyd., 1911, p. 70.

Trophosome.—"Colony (fragmentary) about 3½ inches high, not fascicled, internodes irregular, long. There is but a single dichotomous branching near the top, the usually axillary hydrotheca being present; several of the proximal branches are produced into much annulated shoots, which resemble the so-called stolons found in various groups of hydroids. Hydrothecae enormous for this group, being many times as large as those of *S. polyzonias*, alternate, tubular, doubly curved, the distal extremity being turned slightly upward, about the distal two-thirds free. Margin several times reduplicated, either smooth, or with two or three, sometimes four, inconspicuous teeth. Operculum thick, conspicuous, a single membrane of a single flap where the margin is even, with two flaps where there are two evident teeth, sometimes apparently with more than two flaps, but they are not well defined, probably because the teeth, when three or more, are very low and inconspicuous" (Nutting).

Gonosome.—Unknown.

Distribution.—Off Aleutian island, 283 fathoms (Nutting).

Sertularella minuta Nutting

Plate 35, Fig. 184

Sertularella minuta NUTTING, Am. Hyd., II, 1904, p. 99.

FRASER, West Coast Hyd., 1911, p. 70.

Trophosome.—Colony slender and delicate, attaining a height of about 15 mm.; stem strongly flexuous, divided into regular long internodes with one hydrotheca, and often a branch as well, to each internode; branches irregular, much like the main stem, making a wide angle with the stem; the branches might be considered as regularly dichotomous. Hydrothecae slender, almost tubular, adhering but little to the stem or branch; margin with three teeth; operculum with three flaps.

Gonosome.—Gonangia on stem and branches, ovoid, small as compared with other *Sertularella* gonangia, but not relatively to the size of the colony, with 7 or 8 strong annular ridges; the aperture at the end of a short, tubular neck.

Distribution.—Off Aleutian islands, 283 fathoms (Nutting).

Sertularella pedrensis Torrey

Plate 35, Fig. 185

Sertularella pedrensis TORREY, Hyd. of San Diego, 1904, p. 27.

FRASER, West Coast Hyd., 1911, p. 70.

Vancouver Island Hyd., 1914, p. 191.

Trophosome.—Colony small, stem with occasional small branches or unbranched, divided into rather regular internodes, each of which bears one hydrotheca; hydrothecae alternate, free for over half of their length, narrowing distally; margin with three teeth; operculum with three flaps; surface annulated, annulations most noticeable on the adcauline side.

Gonosome.—Gonangia large, oval, covered throughout, with slender, slightly curved spines.

Distribution.—San Pedro (Torrey); Santa Barbara; Prince Rupert, Clark rock (Fraser); San Diego, San Pedro, Southampton light in San Francisco bay; Nawhitti bar. 8 to 13 fathoms.

Sertularella pinnata Clark

Plate 35, Fig. 186

Sertularella pinnata CLARK, Alaskan Hyd., 1876, p. 226.

NUTTING, Am. Hyd., II, 1904, p. 94.

FRASER, West Coast Hyd., 1911, p. 70.
Vancouver Island Hyd., 1914, p. 191.
Hyd. from west coast of V.I., 1935,
p. 145.
Hyd. Distr. in vicinity of Q.C.I., 1936,
p. 126.

Trophosome.—Colony small but usually growing in dense masses; main stem is divided into short internodes, each of which bears a branch with a hydrotheca in the axil, the branches arranged alternately; the branches often branch again, sometimes dichotomously, giving a flabellate appearance to the whole colony; branches divided into short internodes, each of which bears a hydrotheca; the nodes are deeply cut and besides these, there are often extra constrictions, both on the branch and the hydrotheca; hydrothecae inclined forward and outward so as to appear to be on the front of the stem, distal half or more free; hydrothecae tubular, with a tendency to expansion of the margin; margin with three sharp, strongly marked teeth, two being larger than the third; operculum of three flaps.

Gonosome.—Gonangia borne in two rows on the front of the main stem and branches, oval, strongly rugose, much distorted; collar small, short; aperture circular.

Distribution.—Unalaska, Coal harbour, Shumagin islands, Lituya bay, all in Alaska, 112 fathoms (Clark); San Juan archipelago; Departure bay, Northumberland channel, Dodds narrows, Pylades channel, Friday Harbor, Copalis; Estevan point; Massett harbour (Fraser); San Bruno light in San Francisco bay.   Low tide to 112 fathoms.

Sertularella polyzonias (Linnaeus)

Plate 35, Fig. 187

Sertularia polyzonias LINNAEUS, Syst. Nat., 1758, p. 813.
Sertularella polyzonias NUTTING, Am. Hyd., II, 1904, p. 90.
FRASER, West Coast Hyd., 1911, p. 70.
Vancouver Island Hyd., 1914, p. 191.
Hyd. Distr. in vicinity of Q.C.I., 1936, p. 126.

Trophosome.—Stem rather slender, branching very irregularly, with a tendency to be alternate; branches long and, like the stem, flexuous; the branches may be unbranched or may branch extensively; nodes often appear at regular intervals, each internode bearing a hydrotheca; hydrothecae alternate, rather distant, large, tapering but slightly towards the distal end, the distal half or more free; margin with four teeth that are not very distinct; operculum with four flaps.

Gonosome.—Gonangia large, oval; margin with four distinct, stout spines or teeth; surface strongly but regularly rugose.

Distribution.—Alaska (Clark); Bristol bay, 30 fathoms, north of St. Paul's island, 39-44 fathoms, Alaska (Nutting); San Juan archipelago; Clayoquot sound, Dodds narrows, Gabriola pass, off Matia island, off Waldron island, Friday Harbor; Massett harbour (Fraser); off McArthur reef in Sumner strait, Alaska; off Frederick island, Q.C.I.   15 to 44 fathoms.

Sertularella rugosa (Linnaeus)
Plate 36, Fig. 188

Sertularia rugosa LINNAEUS, Syst. Nat., 1758, p. 813.
Sertularella rugosa NUTTING, Am. Hyd., II, 1904, p. 82.
    FRASER, West Coast Hyd., 1911, p. 70.
        Vancouver Island Hyd., 1914, p. 192.
        Hyd. from west coast of V.I., 1935, p. 145.
        Hyd. Distr. in vicinity of Q.C.I., 1936, p. 126.

Trophosome.—Colony small; stem usually unbranched, divided into regular internodes, each of which bears a hydrotheca; besides the regular nodes, there are several other annulations or constrictions on the stem; hydrothecae alternate, rather distant, fusiform, distinctly and markedly rugose, but the distal fourth or more smooth; margin with four teeth, not very strongly marked; operculum with four flaps.

Gonosome.—Gonangia oval, rugose; margin with four teeth.

Distribution.—Alaska (Clark); Puget sound (Nutting); Popoff island and Yakutat, Alaska; Dodds narrows, Gabriola pass, Friday Harbor; off Long beach, off Sydney inlet; entrance to Flamingo harbour, off Rose harbour in Houston Stewart channel, Massett harbour (Fraser); Shag rock and Lime point in San Francisco bay; Nawhitti bar, off Sadler point, off Klashwan point, 5 miles east of Sand spit, 16 miles northeast of Reef island, off cape James, Hope island; off McArthur reef in Sumner strait, Symonds point in Lynn canal, Alaska.   Low tide to 40 fathoms.

Sertularella tanneri Nutting
Plate 36, Fig. 189

Sertularella tanneri NUTTING, Am. Hyd., II, 1904, p. 81.
FRASER, West Coast Hyd., 1911, p. 70.
Vancouver Island Hyd., 1914, p. 192.
Hyd. Distr. in vicinity of Q.C.I., 1936, p. 126.

Trophosome.—Stem rather slender, not rigid, branching irregularly, the branches having the same appearance as the main stem; stem and branches divided into regular internodes by oblique nodes, above which are annular depressions of the stem; hydrothecae alternate, distant, tubular but narrowed distally, scarcely immersed; margin with four low teeth; operculum of four flaps; surface annulated.

Gonosome.—Gonangium large, oval; surface strongly and regularly rugose; margin with four stout teeth; much like that of *S. polyzonias.* (Gonosome not previously described.)

Distribution.—Off entrance to strait of Fuca (Nutting); Swiftsure shoal; Massett harbour (Fraser); off Virago sound, off Massett inlet, off cape James, Hope island.   8 to 50 fathoms.

Sertularella tenella (Alder)
Plate 36, Fig. 190

Sertularia tenella ALDER, Cat. Zooph. Northumberland, 1857, p. 23.
Sertularella tenella NUTTING, Am. Hyd., II, 1904, p. 83.
FRASER, West Coast Hyd., 1911, p. 70.
Vancouver Island Hyd., 1914, p. 193.
Hyd. Distr. in vicinity of Q.C.I., 1936, p. 126.

Trophosome.—Colony small; branches usually absent; when present, like the main stem; stem slender, geniculate, divided into regular internodes; hydrothecae alternate, one to each node, very distant, scarcely immersed, fusiform; annulated, with the neck smooth; margin with four teeth; operculum with four flaps.

Gonosome.—Gonangia oval or obovate, with seven or eight transverse, annular ridges; distal portion forms a short neck with the circular opening at the end.

Distribution.—Off entrance to strait of Fuca (Nutting); Puget sound (Hartlaub); San Juan archipelago; off Massett, China Hat, Alert bay, Nanoose bay, Departure bay, north of Gabriola island, off Protection island, Northumberland channel, Dodds narrows, off Matia island, Friday Harbor, Swiftsure shoal, Nawhitti bar; off Rose harbour and in the western portion of Houston Stewart channel, Massett harbour (Fraser); San Diego, Wilson's cove, San Clemente, Catalina, Pacific Grove; off Bull harbour, west of Nawhitti bar, off Frederick island, off Sadler point, off Virago sound, off Massett inlet. Low tide to 40 fathoms.

Sertularella tricuspidata (Alder)
Plate 36, Fig. 191

Sertularia tricuspidata ALDER, Ann. and Mag. N.H., (2), XVIII, 1856, p. 356.

Sertularella tricuspidata NUTTING, Am. Hyd., II, 1904, p. 100.

FRASER, West Coast Hyd., 1911, p. 71.

Vancouver Island Hyd., 1914, p. 193.

Hyd. as a food supply, 1933, p. 260.

Hyd. from west coast of V.I., 1935, p. 145.

Hyd. Distr. in vicinity of Q.C.I., 1936, p. 126.

Trophosome.—Stem slender, lax; branching irregularly, usually alternate, but sometimes dichotomous; stem and branches divided into regular internodes, with one hydrotheca or a branch and hydrotheca on each; hydrothecae alternate, distinct, very slightly immersed, tubular, sometimes curved; margin with three distinct teeth; operculum of three flaps; occasionally the hydrothecae are very much prolonged and the margin reduplicated.

Gonosome.—Gonangia numerous on stem and branches; oval, with very strongly crested rugosities; a small tubular neck bears the circular aperture.

Distribution.—Alaska, Aleutian island, St. Paul's island (Clark); Port Townshend (Calkins); Puget sound; off entrance to the strait of Fuca (Nutting); almost everywhere in dredged material in Vancouver island and the Queen Charlotte islands region; in the stomach of *Histrionicus histrionicus* from Unalaska (Fraser); San Diego (Torrey); off the coast of Washington, east of Afognak island, off Chignik bay, off Shumagin islands (Nutting); St. Lawrence island, Bering sea (Jäderholm); Trinity islands, gulf of Alaska (Fraser); in almost every area from Bering sea to San Diego. 1 to 250 fathoms.

Sertularella turgida (Trask)

Plate 36, Fig. 192

Sertularia turgida TRASK, Trans. Cal. Acad. Sc., 1857, p. 113.
Sertularella turgida NUTTING, Am. Hyd., II, 1934, p. 95.
FRASER, West Coast Hyd., 1911, p. 71.
Vancouver Island Hyd., 1914, p. 193.
Hyd. from west coast of V.I., 1935, p. 145.
Hyd. Distr. in vicinity of Q.C.I., 1936, p. 126.

Trophosome.—Colony small; stem stout, either unbranched or with a few irregularly placed branches that are similar to the main stem; hydrothecae alternate, rather distant, nearly tubular but somewhat swollen at the base, more than one-half free; margin with three teeth, two of which are stronger than the third; operculum of three flaps; surface usually smooth.

Gonosome.—Gonangia borne in a row in the axils of the hydrothecae, large, elongated, oval; margin with three or four spines; spines also present in varying numbers on the distal portion of the surface which is not annulated.

Distribution.—Bay of San Francisco, Monterey, Tomales point (Trask); San Diego, Vancouver island (Clark); off Port Simpson (Nutting); San Juan archipelago, Victoria, Port Renfrew, Ucluelet, Dodds narrows, Departure bay; Nawhitti bar, Clayoquot sound, off Lasqueti island, Nanoose bay, Northumberland channel, Pylades channel, Gabriola pass, Gabriola reefs, Porlier pass, off Matia island, Upright channel, Friday Harbor, off Brown island, Port Townshend, Coupeville; in nearly all the shore collections made between Long beach and Esperanza inlet; north of Marble island, Rennell sound; entrance to Big bay, entrance to Flamingo harbour, Danger rocks at the eastern entrance and off Rose harbour in Houston Stewart channel, Massett harbour (Fraser); San Diego, Dillon's beach, Wilson's cove, San Clemente, Catalina, Santa Monica, San Pedro, Whale island, Aumentos rock, Pacific Grove, Trinidad, outside Golden Gate, in all sections of San Francisco bay, Farallone islands, Point Reyes peninsula, Cal.; Newport, off Heceta head, Ore.; off Bull harbour, western portion of Goose island halibut grounds, off Virago sound, 16 miles N.E. of Reef island, cape James, Hope island.   Low tide to 80 fathoms.

Genus **SERTULARIA**

Trophosome.—Hydrothecae in two rows, occurring in pairs which are strictly opposite throughout or at least on the distal portion of the stem or branches.

Gonosome.—Gonangia oval or ovate, usually smooth.

Key to Species

A. Marginal teeth inconspicuous.................*S. desmoides*
B. Marginal teeth distinct
    a. One tooth usually larger than the others.......*S. furcata*
    b. Two teeth, nearly equal
        a. Colony unbranched; gonangium annulated.*S. cornicina*
        b. Colony often with opposite branches; gonangium smooth.........................*S. pumila*

Sertularia cornicina McCrady
Plate 37, Fig. 193

Sertularia cornicina McCrady, Gymno. of Charleston Har., 1858, p. 204.

Nutting, Am. Hyd., II, 1904, p. 30.

Fraser, West Coast Hyd., 1911, p. 72.

Beaufort Hyd., 1912, p. 374.

Trophosome.—Colony in the form of an erect stem, usually less than 15 mm. in height, without branches; stem divided into regular internodes, each of which bears a pair of opposite hydrothecae; hydrothecae tubular, adnate in front for about two-thirds of their length and then turned rather abruptly outwards; margin with two teeth and a two-parted operculum.

Gonosome.—Gonangia borne on the stolon; oval, with a distinct but rather short collar; regularly annulated.

Distribution.—Coronado island, on seaweed at the surface (Torrey); San Pedro, San Pablo point in San Francisco bay.

Sertularia desmoides Torrey
Plate 37, Fig. 194

Sertularia desmoides Torrey, Hyd. of the Pacific coast, 1902, p. 65.

Nutting, Am. Hyd., II, 1904, p. 56.

Fraser, West Coast Hyd., 1911, p. 72.

Trophosome.—Stems little more than 25 mm. in height, appearing regularly on a creeping stolon, slender and flexible; branches scarce and irregular, similar to the main stem, loosely attached to the stem at the base of a pair of hydrothecae; internodes vary in length and in the numbers of pairs of hydrothecae; each pair of hydrothecae is adnate for about half the length on the one side of the stem but they are some distance apart on the other side; the distal half of the hydrotheca turns abruptly almost at right angles to the proximal half; it tapers slightly to the margin, which is faintly two-lipped.

Gonosome.—"Gonothecae borne on stem, sessile, ovate, with a wavy outline and broad, round aperture; half as broad as long. Single gonophore centrally placed, with coenosarcal processes connecting it on all sides with the gonothecal wall" (Torrey).

Distribution.—San Diego, San Clemente island, San Pedro, 1-42 fathoms (Torrey); off San Pedro, 27 fathoms (Nutting); San Diego, Coronado island, San Clemente island, Catalina, San Pedro, point Lobos; Mare island, Southampton light, and Shag rock in San Francisco bay.   1-80 fathoms.

Sertularia furcata Trask
Plate 37, Fig. 195

Sertularia furcata TRASK, Proc. Cal. Acad. Sc., 1857, p. 112.
Sertularia pulchella NUTTING, Am. Hyd., II, 1904, p. 55.
Sertularia furcata FRASER, West Coast Hyd., 1911, p. 72.
> Vancouver Island Hyd., 1914, p. 194.
> Hyd. from west coast of V.I., 1935, p. 145.
> Hyd. Distr. in vicinity of Q.C.I., 1936, p. 126.

Trophosome.—Colony small; stems often in dense masses growing from stolons that are attached to eel-grass or fucus; slender, unbranched, or with few small branches; hydrothecae strictly opposite, each pair, unless two or three near the base, adnate on one side of the stem but some distance apart on the other; they are tubular but somewhat curved and directed well outward, about one-third free; margin with two strong teeth, one of which is usually longer than the other.

Gonosome.—Gonangia borne on the lower portion of the stem, just below the hydrothecae; they are large, oval, somewhat compressed, with a short but distinct collar and large aperture; pedicel short and curved; surface smooth or very slightly wrinkled.

Distribution.—Bay of San Francisco and Farallone islands (Trask); Santa Cruz, Monterey, San Diego, Santa Barbara (Clark); San Pedro, Coronado island, shore to 24 fathoms (Torrey); Ucluelet; Clayoquot sound, Port Grenville; Estevan point, Nootka island, Bajo reef; north shore of Big bay, Q.C.I. (Fraser); San Diego, San Pedro, Trinidad, Aumentos rock, Carmel point, numerous locations outside Golden Gate and in all three sections of San Francisco bay. Low tide to 24 fathoms.

Sertularia gracilis Hincks was included in the 1911 list of west coast hydroids because it was indicated as being found in the Californian region, in Nutting's table of "Geographical Distribution of American Sertularidae", p. 46, American Hydroids, II, 1904, but on page 57, where the description of the species is given, there is no such distribution indicated, nor is there in any of the papers quoted in the synonymy. Since there is a probability that there was a misprint in the table, the description of the species is not included here.

Sertularia pumila Linnaeus
Plate 37, Fig. 196

Sertularia pumila LINNAEUS, Syst. Nat., 1768, p. 807.
  NUTTING, Am. Hyd., II, 1904, p. 51.
  FRASER, West Coast Hyd., 1911, p. 74.
    Can. Atlantic Fauna, 1921, p. 40.

Trophosome.—Stem unbranched or with opposite branches; a pair of hydrothecae to each internode; hydrothecae tubular, free from each other; curved outward, the distal half free.

Gonosome.—Gonangia irregularly arranged on the face of the stem or branches; obovate, smooth or slightly rugose, with a narrow collar and wide aperture.

Distribution.—Coast of California (Clark).

Genus THUIARIA

Trophosome.—Hydrothecae in two rows on stem and branches; not in opposite pairs; hydrothecae with not more than two teeth, operculum of one abcauline flap or two flaps (*T. thuiarioides*, an intergrading form, has an operculum of one adcauline flap).

Gonosome. — Gonangia smooth or with two spines on the shoulders.

Thuiaria alba Fraser
Plate 37, Fig. 197

Thuiaria alba FRASER, West Coast Hyd., 1911, p. 74.
              Vancouver Island Hyd., 1914, p. 196.
              Hyd. Distr. in vicinity of Q.C.I., 1936, p. 126.

Trophosome.—Stem stout, rigid, with several annulations near the base; nodes irregular but distinct; branching regularly pinnate and alternate; the branches are not nearly so stout as the main stem; they are silvery white, while the main stem is much darker, a light horn colour; the hydrothecae are closely crowded, especially on the branches, so much so that in many instances the upper point where the hydrotheca leaves the branch is level with the base of the next hydrotheca in order; those on the two sides alternate regularly; they are tubular, curved so that the margin is nearly vertical; almost wholly immersed; operculum of one abcauline flap.

Gonosome.—Gonangia in two crowded rows on the front of the branches; oval or oblong, with a very short collar; no spines present; there are small annulations in the form of fine lines running transversely.

Distribution.—San Juan archipelago; Alert bay, Friday Harbor, off O'Neale island, San Juan channel, Upright channel, off Brown island, Port Townshend, Griffin bay; off Rose harbour in Houston Stewart channel, 30 fathoms (Fraser).

Thuiaria argentea (Linnaeus)
Plate 37, Fig. 198

Sertularia argentea LINNAEUS, Syst. Nat., 1758, p. 809.
Thuiaria argentea NUTTING, Am. Hyd., II, 1904, p. 71.
FRASER, West Coast Hyd., 1911, p. 75.
Vancouver Island Hyd., 1914, p. 196.
Hyd. from west coast of V.I., 1935, p. 145.
Hyd. Distr. in vicinity of Q.C.I., 1936, p. 126.

Trophosome.—Colonies often growing in clusters, stems slender; branches arise from all sides of the stem but are somewhat scattered; these branch dichotomously but regularly, to produce a graceful colony; the silvery appearance adds to the effect; nodes distant; hydrothecae generally alternate, but occasionally nearly opposite, rather distant, curved gradually outward; usually about one-third free; margin with two teeth, one often longer than the other; operculum of two flaps.

Gonosome.—Gonangia borne on the branches at the base of the hydrothecae, tapering gradually from distal end to proximal, usually with two marked shoulder spines; collar short; opening rather large.

Distribution.—One of the commonest species in shallow water, off the Alaskan coast (Nutting); San Juan archipelago, Queen Charlotte islands; Rose Spit, China Hat, north of Gabriola island, North-umberland channel, Gabriola reefs, off Matia island, off Waldron island, off Shaw island, Friday Harbor, Pacific beach; south of Flores island, off Clayoquot sound; in western portion of Houston Stewart channel, outside Massett inlet (Fraser); Dillon's beach, point Lobos, Bonita point, Lime point, Alcatraz island, Goat island, Southampton light, in San Francisco bay; Petrie point, Humbolt bay, Cal.; western portion of Goose island halibut grounds, off Sad-ler point, off Virago sound. 5 to 65 fathoms.

Thuiaria carica Levinsen
Plate 37, Fig. 199

Thuiaria carica LEVINSEN, Meduser, *etc.*, fra Groenlands Vestkyst, 1893, p. 213.

BROCH, Hydroiden der Arktischen Meere, 1909, p. 176.

FRASER, Monobrachium parasitum, *etc.*, 1918, p. 135.

Trophosome.—Colony consisting of a long and rather rigid stem which is but slightly sinuous where the branches are given off; branches regularly alternating, straight and stiff, either not branched again or having but few short branches resembling the main branches; internodes in the main stem and in the branches varying much in length and in number of hydrothecae given off from each, although the hydrothecae are placed at quite regular intervals. The number of hydrothecae between two branches in succession also varies al-though three is the usual number. There is a distinct process on the stem for the support of each branch, and a distinct joint at the place of attachment. Hydrothecae curved strongly outward so that the margin is vertical; the margin is somewhat sinuous; the abcau-line side does not extend so far outward as the adcauline side.

Gonosome.—(From Broch's description.) The male gonangia appear on the upper branches of the colony where the stem is turned so that the median plane of the hydrothecae is placed almost hori-zontally. They grow upward from the surface of the branch; the point of attachment is just proximal to the base of the hydrotheca, the gonangia alternating from side to side to correspond to the posi-tion of the hydrothecae. The gonangium is obliquely pear-shaped but the margin is at right angles to the axis.

Distribution.—Bull passage, north of Lasqueti island, 25 fathoms (Fraser).

Thuiaria dalli Nutting

Plate 38, Fig. 200

Thuiaria dalli NUTTING, Am. Hyd., 1904, p. 68.

FRASER, West Coast Hyd., 1911, p. 76.

Vancouver Island Hyd., 1914, p. 197.

Hyd. as a food supply, 1933, p. 260.

Hyd. Distr. in vicinity of Q.C.I., 1936, p. 126.

Trophosome.—Stem rather stout, branching regularly; branches alternately arranged, characterized by the fact that they are twisted at the base to be in a plane at right angles to the stem; each is supported on a projection from the stem, consisting usually of two or more joints; hydrothecae nearly opposite, closely placed, almost wholly immersed; margin with two teeth; operculum with two flaps.

Gonosome.—Gonangia of the regular *Thuiaria* type, not very large, obovate, with an aperture of good size, and two shoulder spines, more or less pronounced; each gonangium is inserted without a distinct pedicel, just below the hydrotheca.

Distribution.—Shumagin islands, and Port Moller, Alaska (Clark); Yakutat, Alaska (Nutting); San Juan archipelago, Dodds narrows, Departure bay, Ucluelet; Rose spit, Naden harbour, cape Edenshaw, Nawhitti bar, Claninnick, Nanoose bay, north of Gabriola island, Northumberland channel, Pylades channel, Gabriola pass, Porlier pass, Friday Harbor, Deer harbour, Copalis; Houston Stewart channel in the western portion and off Rose harbour, Massett harbour, outside Massett inlet (Fraser); Jamestown bay and Whale island, Cal.; off Bull harbour, Nawhitti bar, 5 miles east of Sand spit, 16 miles northeast of Reef island; Sitka and Yakutat Alaska. Low tide to 80 fathoms. In the stomachs of *Histrionicus histrionicus* False pass, Alaska, and *Polysticta stelleri* from Izembek bay, Unimak island, Alaska.

Thuiaria distans Fraser

Plate 38, Fig. 201

Thuiaria distans FRASER, Vancouver Island Hyd., 1914, p. 197.

Hyd. from the Queen Charlotte Is., 1936, p. 504.

Trophosome.—Stem erect, geniculate; branching regularly alternate; branches long, slender, geniculate, either unbranched or dichotomously branched; usually three hydrothecae on the stem between two successive branches on the same side; on the branches, the hydrothecae are alternate, very distant for this genus, each hydrotheca placed at a bend in the branch; when the branch is dichotomously branched, there is a hydrotheca in the angle; hydrotheca largest in the centre, tapering both ways, curved to turn well outward, with the margin vertical; about one-half free; margin without distinct teeth, but rather bilateral; operculum of one abcauline flap.

Gonosome.—Gonangia small, elongated, obovate; growing, almost sessile, from the stem and distal branches, the attachment being just below the base of the hydrothecae; opening terminal, with no definite collar.

Distribution.—North of Gabriola island; 9½ miles south of Marble island. 190 fathoms (Fraser).

Thuiaria elegans Kirchenpauer
Plate 38, Fig. 202

Thuiaria elegans KIRCHENPAUER, Nordische Gattungen, 1884, p. 21.
    NUTTING, Am. Hyd., II, 1904, p. 64.
    FRASER, West Coast Hyd., 1911, p. 76.

Trophosome.—Stem slender, erect, reaching a height of 10-11 cm.; somewhat flexuous; divided into irregular internodes by deeply-cut nodes; the upper portion of the colony forms a distinct brush, consisting of the numerous slender, primary and secondary branches; on the remainder of the stem, the branches are broken off, leaving the stumps; the primary and secondary branches, that bear the hydrothecae, are usually flexuous; hydrothecae alternate, adnate throughout; the aperture obliquely cut, so as to form two angles on the otherwise almost horizontal margin; the outer angle is generally more strongly developed, often so strongly that it curves backward to form an inturning hook (after Kirchenpauer).

Gonosome.—Unknown.

Distribution.—Flower bay, Bering sea (Krause).

Thuiaria fabricii (Levinsen)
Plate 38, Fig. 203

Sertularia fabricii LEVINSEN, Vid. Middel. Naturh. Foren., 1892,
    p. 48.
Thuiaria fabricii NUTTING, Am. Hyd., II, 1904, p. 71.

FRASER, West Coast Hyd., 1911, p. 76.
Vancouver Island Hyd., 1914, p. 198.
Hyd. from west coast of V.I., 1935, p.145.
Hyd. Distr. in vicinity of Q.C.I., 1936,
p. 126.

Trophosome.—Stem erect, rather rigid; branches on all sides of the stem, often broken off proximally, leaving a stub in each case; distally forming a dense tuft; the density of the tuft increases by the dichotomous branching which may take place several times; nodes distinct but not regularly placed; hydrothecae usually nearly opposite, but often varying in different parts of the stem and branches; hydrothecae narrowing slightly from base to margin; distal portion free; margin with two teeth; operculum with two flaps.

Gonosome.—Gonangia borne in two rows on the branches, oblong or obovate, with circular aperture and two lateral shoulder spines.

Distribution.—Puget sound (Calkins); Dutch harbour and Orca, Alaska (Nutting); San Juan archipelago and Dodds narrows; Alert bay, Northumberland channel, off Matia island, off Waldron island, off O'Neale island, off Brown island, Deer harbour, Griffin bay, Upright channel, Port Townshend, Ocean beach; off Long beach, off Clayoquot sound; Houston Stewart channel in the western portion and off Rose harbour (Fraser); western portion of Goose island halibut grounds, off Klashwan point, 16 miles northeast of Reef island, off cape James, Hope island; Kadiak and Berg inlet, Alaska. 15 to 50 fathoms.

Thuiaria kurilae Nutting
Plate 38, Fig. 204

Thuiaria kurilae NUTTING, Am. Hyd., II, 1904, p. 65.
FRASER, West Coast Hyd., 1911, p. 76.

Trophosome.—"Specimen about 3 inches high. Stem unbranched, divided into very long and irregular internodes and bearing a row of hydrothecae on each side, there being three hydrothecae, one axillary and two others, between adjacent branches. Branches strictly alternate and divided into long and irregular internodes by distant nodes. Hydrothecae sub-opposite, flask-shaped, the distal end but little constricted. Aperture large, opening outward and a little upward; margin with a very large tooth, a lobe rising upward on the adcauline side and closely appressed to the hydrocaulus" (Nutting).

Gonosome.—Undescribed.

Distribution.—Unalaska (Nutting).

Thuiaria lonchitis (Ellis and Solander)
Plate 39, Fig. 205

Sertularia lonchitis ELLIS and SOLANDER, Nat. Hist. Zooph., 1786, p. 42.

Thuiaria lonchitis NUTTING, Am. Hyd., II, 1904, p. 66.

Trophosome.—Colony reaching a height of 25 to 30 cm.; main stem stout, rigid, straight or slightly flexuous; nodes distinct, often double, but irregularly placed.   Branching irregularly pinnate; the branches stout, stiff, unbranched or but slightly branched; divided into irregular internodes.   Hydrothecae alternate or sub-opposite, tubular, but little tapered, curved outward; about one-fourth free. Margin with two teeth, but these may be not of the same size; operculum of a single abcauline flap.

Gonosome.—"Gonangia borne on upper side of branches, long, slender, with a round aperture, narrow collar and operculum" (Nutting).

Distribution.—Off Frederick island, Q.C.I. 15 fathoms.

Thuiaria plumosa Clark
Plate 39, Fig. 206

Thuiaria plumosa CLARK, Alaskan Hyd., 1876, p. 228.
NUTTING, Am. Hyd., II, 1904, p. 74.
FRASER, West Coast Hyd., 1911, p. 76.

Trophosome.—"Hydrocaulus simple, erect, very slender at the base, increasing in size to the distal end, somewhat twisted, jointed transversely; internodes in the proximal portion of very unequal length, some of them being three times the length of others; those of the upper portion are quite uniform, regularly branched; branches short, arranged alternately, one to an internode, but owing to the twist in the stem, take on a special form, the uppermost erect, lying close to the stem, the lower ones curved outward; attached to the stem by a very prominent process, bearing a few branchlets, regularly jointed; branchlets do not extend beyond the end of the branches, and lie close to the latter.   Hydrothecae largest at the base, tapering slightly outward, entirely immersed, aperture towards the stem, the outer side produced, rim ornamented with two large teeth placed at the outer side; two tooth-like processes of the perisarc also occur at the base of each hydrotheca; arranged alternately on the branches and branchlets; upon the stem there are three to each internode, two placed opposite to each other and one in the axil of the branch" (Clark).

Gonosome.—"Gonangia sessile, very long and narrow, tapering very gradually to the base, ornamented with two short horns placed on opposite sides of the orifice, near the distal end. Orifice terminal, large; borne in single rows on the upper side of the branches and branchlets" (Clark).

Distribution.—Nunivak island, Bering sea, 30 fathoms (Clark); Bering sea (Jäderholm).

Thuiaria robusta Clark
Plate 39, Fig. 207

Thuiaria robusta CLARK, Alaskan Hyd., 1876, p. 227.
NUTTING, Am. Hyd., II, 1904, p. 64.
FRASER, West Coast Hyd., 1911, p. 76.
Vancouver Island Hyd., 1914, p. 198.
Hyd. as a food supply, 1933, p. 260.

Trophosome.—Stem stout with deeply cut internodes; branches also stout arising from all sides of the stem, the proximal often broken off and the distal forming a dense tuft; hydrothecae alternately, not closely placed; proximally almost wholly immersed, distally with a small portion free; long, tubular, very slightly larger at the base than at the margin; margin bi-labiate; operculum with two flaps on the distal hydrothecae and one flap on the proximal.

Gonosome.—"Gonangia borne in rows on the terminal branchlets; slender, with a terminal collar and aperture, and two long curved spines rising from antero-lateral corners of the shoulders" (Nutting).

Distribution.—Hagmeister island and King's island, Bering sea (Clark); entrance to strait of Fuca, off Sitka, north of Unimak pass, south of Nunivak island, Bering sea, south of Pribilof islands, 13 to 51 fathoms (Nutting); in the stomach of *Polysticta stelleri*, Izembek bay, Unimak island, Alaska; Symonds point in Lynn canal, Alaska. 9 fathoms.

Thuiaria similis (Clark)
Plate 39, Fig. 208

Sertularia similis CLARK, Alaskan Hyd., 1876, p. 219.
Thuiaria similis NUTTING, Am. Hyd., II, 1904, p. 69.
FRASER, West Coast Hyd., 1911, p. 77.
Vancouver Island Hyd., 1914, p. 199.
Hyd. Distr. in vicinity of Q.C.I., 1936, p. 126.

Trophosome.—Colony bilateral, with the main stem very distinct and much stouter than the branches; branching regularly alternate; hydrothecae usually in nearly opposite pairs which vary much in the distance from one to the other, although the distance may be fairly constant in the same colony; hydrothecae slender, tubular, tapering but slightly to the margin, distal portion free and turned well outward; sometimes they are much prolonged, in which case they project for a large portion of their length; margin with two distinct teeth; operculum with two flaps.

Gonosome.—Gonangia borne singly on the branches, oval, with a short, stout collar and wide aperture; no very distinct pedicel; surface free of annulations or spines.

Distribution.—Hagmeister island, Bering sea (Clark); Puget sound, off the entrance to strait off Fuca, Berg inlet, Glacier bay, south of Unimak pass, south of Pribilof islands, west of Unimak island, 13-72 fathoms (Nutting); Bare island (Hartlaub); San Juan archipelago, Dodds narrows, Departure bay; Rose spit, Massett, China Hat, Claninnick, Port Renfrew, Nawhitti bar, north of Gabriola island, Northumberland channel, Pylades channel, Ruxton passage, Gabriola pass, Gabriola reefs, off Matia island, off O'Neale island, off Shaw island, off Brown island, San Juan channel, Deer harbour, Upright channel, Griffin bay, Port Townshend; Houston Stewart channel, in the western portion and off Rose harbour (Fraser); Angel island, Goat island, Southampton light, San Pablo bay, all in San Francisco bay; off Heceta head, Ore.; western portion of Goose island halibut ground, Kaison bank, off Frederick island, off Klashwan point, east of Rose spit, 16 miles northeast of Reef island; Kadiak, point Gardiner buoy off Admiralty island, Alaska. 3 to 72 fathoms.

Thuiaria tenera (Sars)
Plate 39, Fig. 209

Sertularia tenera SARS, Bidrag til Kundskaben, 1873, p. 20.
Thuiaria tenera NUTTING, Am. Hyd., II, 1904, p. 70.
FRASER, West Coast Hyd., 1911, p. 78.
Vancouver Island Hyd., 1914, p. 200.

Trophosome.—Stem rather slender; branches arising from opposite sides of the stem but somewhat scattered; these branch dichotomously and regularly; hydrothecae alternate, rather distant, enlarged above the base, then narrowing rapidly to the margin; fully one-half of the hydrotheca is free and projects well outwards but the margin is directed upward; margin usually with two large blunt teeth but these may be so low as not to be readily noticeable; operculum of two flaps or one abcauline flap.

Gonosome.—Gonangia borne singly on the branches, oval, with a short, stout collar and wide circular aperture; no very distinct pedicel; surface free of annulations or spines.

Distribution.—Off entrance to strait of Fuca, Kadiak island, St. Paul's island, Bering strait, Alaska (Nutting).   40 fathoms.

Thuiaria thuiarioides (Clark)
Plate 39, Fig. 210

Sertularia thuiarioides CLARK, Alaskan Hyd., 1876, p. 223.
Thuiaria thuiarioides NUTTING, Am. Hyd., II, 1904, p. 64.
                FRASER, West Coast Hyd., 1911, p. 78.
                Vancouver Island Hyd., 1914, p. 200.
                Hyd. as a food supply, 1933, p. 260.
                Hyd. from the west coast of V.I., 1935, p. 145.
                Hyd. Distr. in vicinity of Q.C.I., 1936, p. 126.

Trophosome.—Main stem rather stout; branches from all sides of the stem, placed closely enough to make a dense tuft; branches branch dichotomously; hydrothecae as nearly being opposite as distinctly alternate; tubular below, narrowing towards the margin, a small portion free; margin circular, facing upwards; operculum of one adcauline flap.

Gonosome.—Gonangia oblong or obovate, with a large circular aperture; two small shoulder spines.

Distribution.—Bering sea, west of Nunivak island, Chignik bay, Alaska (Clark); Puget sound (Calkins); Rose spit, Clayoquot sound, Swiftsure shoal, Copalis; off Sydney inlet; off Rose harbour in Houston Stewart channel, Massett harbour, off Massett inlet in the stomach of *Somateria spectabilis* and *Polysticta stelleri* in Izembek bay, Unimak island (Fraser); Yakutat, off St. Lawrence island, Bering sea (Nutting).   13 to 30 fathoms.

Thuiaria thuja (Linnaeus)
Plate 40, Fig. 211

Sertularia thuja LINNAEUS, Syst. Nat., 1758, p. 809.
Thuiaria thuja NUTTING, Am. Hyd., II, 1904, p. 62.
                FRASER, West Coast Hyd., 1911, p. 78.
                Vancouver Island Hyd., 1914, p. 201.

Trophosome.—Main stem rigid, not very stout; branches from all sides of the stem, proximal ones usually broken off, leaving a stub; all branches stiff, branching dichotomously several times, making a dense tuft, often spoken of as the "bottle-brush"; hydrothecae alternate, closely placed, almost wholly immersed, tubular; margin almost vertical, without teeth; operculum single abcauline flap.

Gonosome.—Gonangia in rows that may be crowded on the stem and proximal portions of the branches; oval, with short collar and large terminal aperture and a short, distinct pedicel; surface without annulations or spines.

Distribution.—Bering sea (Stimpson); south of Unimak pass, 45 fathoms, south of Pribilof islands, 25 fathoms (Nutting); San Juan archipelago, Banks island; off Matia island, off Waldron island, Friday Harbor (Fraser).

Family **Plumularidae**

Trophosome.—Hydrothecae growing only on one side of the branches (hydrocladia), sessile, more or less adnate; nematophores always present.

Gonosome.—Gonophores producing fixed sporosacs, which are often protected by special modifications of the branches.

Key to Genera

A. Statoplaean forms, *i.e.*, those with fixed nematophores that are usually monothalamic
    a. Gonangia well protected
        a. Gonangia protected by corbulae.........*Aglaophenia*
        b. Gonangia protected by phylactogonia....*Cladocarpus*
    b. Gonangia unprotected
        a. Hydrothecal ridges or septa present
            1. Stem simple, hydrocladia unbranched.*Diplocheilus*
            2. Stem fascicled, hydrocladia branched...*Nuditheca*
        b. Hydrothecal ridges or septa absent........*Tetranema*
B. Eleutheroplaean forms, *i.e.*, those with movablenema tophores that are usually bithalamic
    a. Hydrocladia growing directly from the stolon..*Antennella*
    b. Hydrocladia growing from the stem
        a. Hydrocladia pinnately arranged.........*Plumularia*
        b. Hydrocladia in whorls or scattered over the stem.....
        ....*Antennularia*

Genus AGLAOPHENIA

Trophosome.—Hydrothecal margin provided with sharp teeth; posterior intrathecal ridge present; one mesial and two supracalycine nematophores for each hydrotheca always present.

Gonosome.—Gonangia enclosed in true corbulae, formed of modified pinnae. There are no hydrothecae at the base of the gonangial leaves.

KEY TO SPECIES

A. Hydrothecal margin with 9 teeth
   a. Median tooth retrorse
      a. Corbula long, with 8 pairs of leaves........*A. diegensis*
      b. Corbula short, with 4-6 pairs of leaves..*A. inconspicua*
   b. Median tooth not retrorse
      a. Corbula with 9 pairs of leaves, hydrotheca sigmoid....
      ....*A. pluma*
      b. Corbula with 10 pairs of leaves, hydrotheca straight...
....*A. lophocarpa*

B. Hydrothecal margin with 11 teeth
   a. Corbula with 8 pairs of leaves
      a. One hydrotheca on the hydrocladium between corbula and stem, mesial nematophore narrowing towards tip.............................*A. octocarpa*
      b. Two hydrothecae between corbula and stem, mesial nematophore expanding towards tip..*A. latirostris*
   b. Corbula with 13 pairs of leaves, 3 hydrothecae between corbula and stem.................*A. struthionides*

Aglaophenia diegensis Torrey
Plate 40, Fig. 212

Aglaophenia diegensis TORREY, Hyd. of the Pacific coast, 1904, p. 71.
    FRASER, West Coast Hyd., 1911, p. 80.

Trophosome.—Colonies large for this genus, reaching a height of 15 cm.; stem unbranched, divided into regular short internodes, each bearing a hydrocladium; hydrocladia regularly alternate, but those of the two sides are not in the same plane; hydrocladia divided into regular, short internodes by slightly oblique nodes; one adnate hydrotheca to each internode; hydrothecae slightly longer than broad, the margin with 9 irregular teeth; the median tooth is sharp and slightly recurved; the teeth next the median are the longest, they are directed forward; the second and third are nearly equal and the fourth but slightly smaller; the intrathecal ridge is not very prominent; supracalycine nematophores reach the margin of the hydrotheca; the median is large and reached to, or almost to, the margin; the cauline nematophore, at the base of the hydrocladium, is triangular; two others, tubular, are present on each internode of the stem.

Gonosome.—Corbulae about three times as long as broad, usually but one hydrotheca between the corbula and the axil of the hydrocladium. There are 8 pairs of leaves, each with a row of nematophores on the margin; no pronounced processes at the base of the leaves.

Distribution.—San Diego and False bay, Cal., 1-7 fathoms (Torrey); San Diego; Mile rock, San Bruno light, and Shag rock in San Francisco bay. Low tide to 7 fathoms.

Aglaophenia inconspicua Torrey
Plate 40, Fig. 213

Aglaophenia inconspicua TORREY, Hyd. of the Pacific coast, 1904, p. 71.

FRASER, West Coast Hyd., 1911, p. 80.

Trophosome.—Stem stout, 35-40 mm. high, divided into regular internodes by oblique nodes; hydrocladia regularly alternate but those on the two sides are not in the same plane; short, divided into regular internodes, by well marked, transverse nodes; one hydrotheca to each node, the distal fourth free and projecting well out from the hydrocladium; length and breadth nearly equal; the margin with 9 teeth; the median sharp and retrorse; the next pair large, projecting outward; the second pair smaller than either the first or third pair; the fourth pair slender; intrathecal ridge well marked; supracalycine nematophores short, not reaching the margin of the hydrotheca; the mesial large and usually reaching the margin; the three cauline nematophores are similar to the supracalycine, but the axillary is smaller than either of the other two.

Gonosome.—Corbulae short and deep, arched; one hydrotheca between the corbula and the axil of the hydrocladium; 4 to 6 leaves with nematophores on the margin; these have a margin making a sharper angle with the sides than usual in nematophores in this position. No pronounced processes at the base of the leaves.

Distribution.—San Diego, 5 fathoms (Torrey); San Diego; off Mile rock, Mare island light, and Shag rock in San Francisco bay; off Heceta head, Ore.   5 to 84 fathoms.

Aglaophenia latirostris Nutting
Plate 40, Fig. 214

Aglaophenia latirostris NUTTING, Am. Hyd., I, 1900, p. 101.

FRASER, West Coast Hyd., 1911, p. 80.

Vancouver Island Hyd., 1914, p. 202.

Hyd. Distr. in vicinity of Q.C.I., 1936, p. 126.

Trophosome.—Colony unbranched, attaining a height of about 5 cm.; stem simple, divided into regular internodes, each of which bears a hydrocladium; hydrocladia lying in the same plane, alternate, closely approximated; hydrocladia divided into internodes, that are almost as broad as long, each internode bearing a hydrotheca.   Hydrothecae closely approximated, rapidly increasing in diameter from base to margin; margin with 11 teeth; the median retrorse; the next pair projecting forward and the remaining four pairs erect but somewhat irregular; intrathecal ridge well marked; mesial nematophore large, adnate to the hydrotheca almost to its margin and then projecting forward much farther than the margin of the hydrotheca, with a spout-like distal extremity; supracalycine nematophores small, not reaching the margin of the hydrotheca.

Gonosome.—"Corbula closed, composed of about eight pairs of moderately narrow leaves, each of which bears a row of nematophores on its distal edge and another on its inner, proximal edge, as in *A. struthionides.*   There is an aperture between the bases of adjacent leaves, and no prominent spur at the bases.   There are two hydrothecae between the corbulae and the stem" (Nutting).

Distribution.—Santa Barbara, off Oregon coast, Puget sound; Massett harbour (Fraser); off Brothers light, Southampton light, and Mile rock in San Francisco bay; Massett harbour.   7 to 16 fathoms.

Aglaophenia lophocarpa Allman
Plate 40, Fig. 215

Aglaophenia lophocarpa ALLMAN, Mem. Mus. Comp. Zool., 1877, p. 101.

NUTTING, Am. Hyd., I, 1900, p. 92.

STECHOW, Zool. Jahrb., 47, 1923, p. 250.

Trophosome.—"Hydrocaulus attaining a height of between two and three inches, simple, not fascicled; pinnae alternate, springing from a point near the distal end of each internode. Hydrothecae deep, somewhat tumid below, margin slightly everted, with nine equal, very distinct teeth; intrathecal ridge transverse. Supracalycine nematophores slightly overtopping the hydrothecae; mesial nematophore adnate to within a very short distance of the summit and attaining nearly half of the height of the hydrotheca; cauline nematophores two on each internode of main stem, one close to the axil of the pinna and the other near the proximal end of the internode" (Allman).

Gonosome.—"Corbulae with about ten pairs of leaflets; leaflets broad, united into a completely closed corbula, the distal margin of each carrying numerous well-developed denticles, and projecting from the sides of the corbula in the form of a pectinated ridge which is continued as a free serrate crest beyond the roof; a spur-like denticle at the base of each leaflet; peduncle of corbula carrying a single hydrotheca" (Allman).

Distribution.—Pacific Grove (Stechow).

Aglaophenia octocarpa Nutting
Plate 41, Fig. 216

Aglaophenia octocarpa NUTTING, Am. Hyd., I, 1900, p. 103.
FRASER, West Coast Hyd., 1911, p. 80.

Trophosome.—Colony reaching a height of about 5 cm.; stem rather slender, divided into short internodes by transverse nodes, that are not always distinct. Hydrocladia regularly alternate, those on the two sides approaching each other on the face of the stem, often attached nearer the base of the internode than the distal end, long and curved; divided into regular internodes by distinct nodes; one hydrotheca to each internode, broader at the margin than at the base; adnate for about two-thirds of its length, the free portion projecting very little; margin with eleven teeth, the median one being retrorse, the next pair prominent and projecting outward, the others somewhat irregular but approximately equal. Besides the posterior intrathecal ridge there is a very pronounced anterior ridge, extending nearly across the hydrotheca; supracalycine nematophores somewhat obliquely placed, the margin reaching the level of the margin of the hydrotheca; mesial nematophore prominent, projecting well outward from the hydrotheca and reaching to, or almost to, the level of the hydrothecal margin; the three cauline nematophores prominent, alike in size and shape.

Gonosome.—Corbula twice as long as deep, with one hydrotheca between the corbula and the axil of the hydrocladium; eight pairs of leaves with strong nematophores on the margin and a short backwardly directed process at the base of each.

Distribution.—Off Santa Barbara, 238 fathoms, off Santa Rosa island, 41 fathoms.

Aglaophenia pluma (Linnaeus)
Plate 41, Fig. 217

Sertularia pluma LINNAEUS, Syst. Nat., 1767, p. 1309.
Aglaophenia pluma TORREY, Hyd. of the Pacific coast, 1902, p. 73.
          FRASER, West Coast Hyd., 1911, p. 80.

Trophosome.—Colony small, reaching not more than 35 mm. in height; stem dark brown, slender curved, divided by transverse nodes into regular internodes, much longer than broad; hydrocladia light in colour, regularly alternate, those on the two sides approaching each other on the face of the stem, each hydrocladium supported on a projecting shoulder of the internode; each hydrocladium divided into regular internodes by nodes that are not always distinct; one hydrotheca to each internode, adnate for only about one-third of the length, slightly sigmoid in shape.   Margin with 9 regular teeth; unlike that in most of the species on the coast, the median tooth is not retrorse, but projects slightly outward; the intrathecal ridge is well marked.   Supracalycine nematophores more evident than usual on account of the shape of the hydrotheca, reaching to the margin of the hydrotheca; the mesial, short, reaching little more than half the length of the hydrotheca, projecting very little; there are three cauline nematophores, one small one, near the centre of the internode, another small one at the base of the hydrocladium shoulder, on the outside, and a much larger one, near the end of the shoulder, but on the side next the stem.

Gonosome.—Corbula long and slender, nearly three times as long as deep, with one hydrotheca between the corbula and the axil of the hydrocladium; nine pairs of leaves with nematophores on the margin but no prominent process at the base; the margin of the leaves appears thickened and is much darker than the remainder of the leaves.

Distribution.—Off Coronado island, on kelp (Torrey); Coronado island, off Bajo reef, 23 fathoms.

Aglaophenia struthionides (Murray)
Plate 41, Fig. 218

Plumularia struthionides MURRAY, Ann. and Mag. N.H., (3), V, 1860, p. 251.

Aglaophenia struthionides NUTTING, Am. Hyd., I, 1900, p. 102.
FRASER, West Coast Hyd., 1911, p. 80.
Vancouver Island Hyd., 1914, p. 203.
Hyd. as a food supply, 1933, p. 260.
Hyd. from west coast of V.I., 1935, p. 145.
Hyd. Distr. in vicinity of Q.C.I., 1936, p. 126.

Trophosome.—Colonies usually growing in large, conspicuous bunches, which appear in view on the rocks at low tide. The stems are usually unbranched but occasionally scattered branches are given off, making a wide angle with the stem and apparently loosely connected with it; the branch has the same appearance as the main stem; the stem is divided into regular short internodes, each of which bears a hydrocladium; the hydrocladia are regularly alternate but those on the two sides do not come out in the same plane; the hydrocladia are directed outward and slightly upward; they are regularly graded in length to give the "ostrich-plume" effect; they are divided into regular internodes, each of which bears a hydrotheca, which occupies practically all the one side, hence the hydrothecae are closely approximated. Each hydrotheca has eleven teeth; the median one is retrorse, the one on each side of it is pointing forward; these three are quite sharp; the remaining four on each side are much blunter and are quite irregular; the intrathecal ridge is evident; the supracalycine nematophores are large but do not overtop the hydrotheca; the mesial nematophore is large, the free portion varying much in length, but seldom reaching beyond the margin of the hydrotheca; the cauline nematophore at the base of the hydrocladium is large and triangular.

Gonosome.—Corbula about three times as long as deep, with three hydrothecae between the corbula and the axil of the hydrocladium; with thirteen pairs of leaves, each with a row of nematophores showing along the margin; no pronounced process at the base of the leaves.

Distribution.—Santa Cruz (Nutting); San Diego (Palmer); San Francisco (A. Agassiz); Townshend bay (Calkins); Puget sound to San Diego (Torrey); widely distributed at low tide and in shallow water in the whole Queen Charlotte island and Vancouver island region (Fraser); in the stomach of *Histrionicus histrionicus* in Barkley sound (Fraser); San Diego, Santa Monica, 3 locations off Santa Cruz, off point Pinos, San Pedro, Pacific Grove, Pillar point, San Mateo, 105 locations in San Francisco bay region, Point Reyes, Cal.; off Heceta head, Newport, Ore.; further records in various portions of the coast of British Columbia; off McArthur reef in Sumner strait, Alaska. Low tide to 85 fathoms.

Genus ANTENNELLA

Trophosome.—"Colony consisting of hydrocladia springing directly from the hydrorhiza without the intervention of stem or branches; hydrocladial internodes and hydrothecae as in the *catharina* group of the genus *Plumularia*" (Nutting).

Gonosome.—"Gonothecae ovate, unprotected" (Torrey).

Antennella avalonia Torrey
Plate 41, Fig. 219

Antennella avalonia TORREY, Hyd. of the Pacific coast, 1902, p. 74.
FRASER, West Coast Hyd., 1911, p. 81.

Trophosome.—"Stems rooted by creeping stolon, unbranched; largest 7 mm. high, with five hydrothecae. Each stem divided by alternately oblique and transverse nodes, which are always weak, into alternating thecate and intermediate internodes. Hydrothecae as deep as broad, free for half their length, with slightly everted, circular margin. Mesial nematophores borne on small processes of the stem. Intermediate internodes with one or two nematophores, never three" (Torrey).

Gonosome.—"Gonothecae broadly ovate, with short, slightly ringed peduncles, borne in pairs on the thecate internodes, one on each side of the mesial nematophore. Pair of nematophores at the base of each" (Torrey).

Distribution.—Avalon, Catalina island (Torrey).

Genus ANTENNULARIA

Trophosome.—Hydrocladia arranged in whorls or scattered over the stem.

Gonosome.—Gonangia unprotected.

Antennularia verticillata Fraser
Plate 41, Fig. 220

Antennularia verticillata FRASER, New and unreported hyd., 1925, p. 171.

Trophosome.—A fragment of a stem 2.5 cm. long is stout and uniform, diameter 1.0 mm. The canaliculated appearance of the coenosarc is very regular, there being 24 vertical, parallel grooves showing at the surface; hydrocladia arranged in very regular whorls of six, the individual hydrocladia in one whorl opposite the spaces between the hydrocladia in the whorl above and the whorl below, so that there are really twelve vertical series, and thus two coenosarcal canals for each series; the basal hydrocladial internode is long, curved, and so much stouter than the remainder of the hydrocladium that it might be considered a process of the stem, bearing the hydrocladium, particularly since it is at the distal extremity of the internode that the hydrocladium breaks off the most readily; this basal internode bears two nematophores; next to it, there is a hydrothecate internode, with one nematophore below the hydrotheca and two above it; a non-hydrothecate internode with two nematophores follows; the hydrothecate and non-hydrothecate internodes then alternate throughout the hydrocladium.

Gonosome.—Unknown.

Distribution.—31.7 miles N. 75°E. of Heceta head light, Ore. 84 fathoms (Fraser).

Genus CLADOCARPUS

Trophosome.—Hydrothecae deep with the margin smooth or with low, blunt teeth; mesial nematophore short.

Gonosome.—Gonangia borne on the stem, at the base of the hydrocladia, protected by processes (phylactogonia) springing from the base of the hydrocladia; these have nematophores but no hydrothecae.

Cladocarpus vancouverensis Fraser
Plate 41, Fig. 221

Cladocarpus vancouverensis FRASER, Hyd. of Vancouver Island region, 1914, p. 204.

Trophosome.—Stems simple, unbranched, longest specimen 12 cm.; hydrocladia regularly alternate, those on the two sides not in the same plane, divided into regular internodes; hydrothecae much deeper than wide, tapering slightly but gradually from base to margin, adnate throughout; margin with one central, distinct, but not large, sharp tooth, the remainder weakly crenulated. The supracalycine nematophores are long but do not reach beyond the margin of the hydrotheca; the mesial nematophore is projected outward, distal portion free, jointed near the base; a septal ridge is present at the base of the supracalycine nematophore, one at the base of the hydrotheca and two others regularly placed between these.

Gonosome.—Gonangia borne on the front of the stem and protected by phylactogonia, which are two-pronged, but each of these prongs may be two-pronged; they are oval or somewhat ovate, with distal end rounded.

Distribution.—Lasqueti island, West rocks, Northumberland channel (Fraser); off Long point, Catalina island; off Virago sound, off Massett inlet. 15-178 fathoms.

Genus DIPLOCHEILUS

Trophosome.—"All internodes thecate, each internode with an infracalycine, mesial nematophore, not in contact with the hydrotheca, and a supracalycine, median sarcostyle without definite nematophores; each hydrotheca with anterior intrathecal ridge" (Torrey).

Gonosome.—"Gonangia unprotected" (Torrey).

Diplocheilus allmani Torrey
Plate 42, Fig. 222

Halicornaria producta TORREY, Hyd. of the Pacific coast, 1902, p. 75.

Diplocheilus allmani TORREY, Hyd. of San Diego, 1904, p. 36.

FRASER, West Coast Hyd., 1911, p. 81.

Trophosome.—"Colony with simple stem, divided obliquely into internodes which vary in length according to age. Hydrothecae alternate, each from a shoulder process projecting from the middle region of each internode. Each hydrocladium divided more or less obliquely into equal thecate internodes. Each hydrotheca somewhat compressed below, somewhat flaring distally, with a broadly oval, smooth orifice; about as deep as long; free for one-third of its length; strong anterior intrathecal septum about two-thirds of the length of the hydrotheca from the bottom; reaching about one-third across it at widest point. Cauline nematophores absent, with the exception of single axillary nematophores. Mesial nematophore short, not reaching the base of the hydrotheca, expanding into the form of a sickle-shaped segment of a saucer, with a diameter two-thirds that of the hydrotheca and embracing the internode for half of its circumference. Single median supracalycine sarcostyle, flanked by two webs of perisarc stretched between theca and internode, forming a non-typical median nematophore" (Torrey).

Gonosome.—Unknown.

Distribution.—San Diego and Point Loma, Cal., along shore, growing on seaweed (Torrey).

Genus NUDITHECA

Trophosome.—"Stem fascicled; hydrocladia compound or branched; supracalycine and mesial nematophores present; hydrothecal margin without teeth" (Nutting).

Gonosome.—"Gonangia borne singly on the hydrocladia, and devoid of phylactogonia, but with two or three nematophores on their pedicels" (Nutting).

Nudittheca dalli (Clark)
Plate 42, Fig. 223

Macrorhynchia dalli CLARK, Alaskan Hyd., 1876, p. 230.
Nuditheca dalli NUTTING, Am. Hyd., I, 1900, p. 129.
FRASER, West Coast Hyd., 1911, p. 81.

Trophosome.—"Colony branched, attaining a height of 5 inches; stem coarse, strongly fascicled; hydrocladia closely approximated, compound; consisting of a main, straight branch, which usually gives off three branchlets from its proximal portion; main branch hydrothecate, except in the region from which the branchlets originate; branchlets regularly hydrothecate, with a hydrotheca in the axil of each; hydrocladia divided into short internodes, each with a strong internal septal ridge, opposite the base of the hydrotheca and another opposite the supracalycine nematophores; proximal portion of each internode very broad, forming a shoulder on its front side, upon which the hydrotheca rests. Hydrothecae broad, cup-shaped margin slightly expanded and smooth; no intrathecal ridge; supracalycine nematophores broad, somewhat expanded above, and with a strong internal ridge near the base; mesial nematophores resting on the broadened base of the internode, short and free, slightly expanded above. There are two or three nematophores on each internode of that portion of the main branch of the hydrocladium which bears the branchlets; cauline nematophores numerous" (Nutting).

Gonosome.—"Gonangia very large, long, almost cylindrical, borne on the branchlets of the hydrocladia on the distal part of the colony. There are two or three nematophores near the base of each gonangium" (Nutting).

Distribution.—Unalaska and Akutan pass, Alaska, on the beach (Clark); Unalaska (Stechow).

Genus PLUMULARIA

Trophosome.—Hydrocladia usually unbranched, pinnately arranged, each ordinarily having more than one hydrotheca; hydrothecae with entire margin; all nematophores movable.

Gonosome.—Gonangia without extra protection.

KEY TO SPECIES

A. Stem fascicled
    a. Hydrothecate and non-hydrothecate internodes alternating on the hydrocladiun; no septal ridges...*P. halecioides*
B. Stem simple
    a. Hydrothecate and non-hydrothecate internodes alternating in the hydrocladium
        a. Hydrocladial internodes with strong septal ridges
            1. Hydrocladial nodes alternately transverse and oblique......................*P. alicia*
            2. Hydrocladial nodes all transverse
                i. Colony much branched.........*P. corrugata*
                ii. Colony unbranched or but slightly branched.. ....*P. lagenifera*
        b. Septal ridges absent or indistinct
            1. Non-hydrothecate internode short, with one nematophore ....*P. setacea*
            2. Non-hydrothecate internode longer, with two nematophores.............*P. megalocephala*
    b. All hydrocladial internodes hydrothecate
        a. Hydrocladial internodes with strong septal ridges.....
        ....*P. virginiae*
        b. Septal ridges absent or indistinct
            1. Only one hydrocladium to a cauline internode....
            ....*P. plumularoides*
            2. Commonly two hydrocladia to a cauline internode, particularly in the proximal portion..*P. goodei*

Plumularia alicia Torrey
Plate 42, Fig. 224

Plumularia alicia TORREY, Hyd. of the Pacific coast, 1902, p. 75.

Hyd. of San Diego, 1904, p. 37.

FRASER, West Coast Hyd., 1911, p. 82.

Trophosome.—Stems slender, unbranched or loosely branching; divided into regular internodes by transverse nodes. Hydrocladia arranged alternately, each from a distinct shoulder near the distal end of the internode. Each hydrocladium is divided into internodes by alternating transverse and oblique nodes; the internodes are alternately non-thecate and thecate, the basal internode being non-thecate; the non-thecate internode is much shorter than the thecate, with two septal ridges dividing the internode into three nearly equal portions; the thecate internodes have two septal ridges near the extremities. The hydrotheca is approximately as deep as it is broad, placed on a definite shoulder on the proximal portion of the internode; the inner wall is free for the greater portion of its length; the margin, which makes an angle of about 45° with the hydrocladium, is oval and even. A single nematophore is present on each cauline internode, opposite the hydrocladial shoulder and there are two nematophores in the axil of the hydrocladium; one nematophore on each non-thecate internode, one on the shoulder of the thecate internode and one above the hydrotheca.

Gonosome.—"Male gonophores small, ovate, attached by very short peduncles between the nematophores in the axils of the stem or branches, one to an axil; chitinous investment very thin"(Torrey).

Distribution.—San Diego and Long Beach, Cal., 5-25 fathoms (Torrey); San Diego, Long beach, Carquinez light in San Francisco bay; off Heceta head, Ore.   13-180 fathoms.

Plumularia corrugata Nutting
Plate 42, Fig. 225

Plumularia corrugata NUTTING, Am. Hyd., I, 1900, p. 64.

FRASER, West Coast Hyd., 1911, p. 82.

Vancouver Island Hyd., 1914, p. 205.

Hyd. from west coast of V.I., 1935, p. 145.

Trophosome.—Stem simple, erect, with irregular annulations at the base; divided into regular internodes, each of which gives off a hydrocladium from a process at its distal end; hydrocladia alternating, lying in the same plane, the proximal ones unbranched but the distal usually giving off several branches; hydrocladia slender, consisting of alternating thecate and non-thecate internodes; the proximal internode is short, with one septal ridge, non-thecate; the second, with usually four well pronounced ridges, bears a hydrotheca, which is about as deep as broad, placed slightly distal to the centre; the third has two septal ridges, and the remainder of the hydrocladium consists of internodes like the second and third, alternating; the intermediate internode is somewhat shorter than the thecate but each is rather long and slender; there are two supracalycine nematophores, a mesial on each hydrocladial internode with the exception of the first, one on each cauline internode on the side opposite the process for the hydrocladium and one in the axil of each hydrocladium.

Gonosome.—Gonangia of two kinds,—one nearly oval, about three times as long as broad, with a truncated top or the appearance of a slight collar and the other elongated, with or without a bottle neck.   In both cases they are found attached to the process which supports the hydrocladium.

Distribution.—San Juan archipelago; Nanoose bay, Departure bay, Clarke rock, north of Gabriola island, Protection island, Gabriola pass, off Matia island; off Clayoquot sound (Fraser); western portion of Houston Stewart channel, off Massett inlet.   30 to 45 fathoms.

Plumularia goodei Nutting
Plate 43, Fig. 226

Plumularia goodei NUTTING, Am. Hyd., I, 1900, p. 64.
FRASER, West Coast Hyd., 1911, p. 82.
Vancouver Island Hyd., 1914, p. 206.

Trophosome.—Colony small, less than 25 mm. in height; stems growing from a coarse network of stolons; stem simple, divided into regular internodes, each of which may bear one or two, possibly three, unbranched hydrocladia.   Commonly the proximal internode bears two hydrocladia, while the distal bear one each; in either case, the hydrothecae are alternate; there may be one or two non-thecate internodes at the base of the hydrocladium, or these may be absent; all the rest of them bear one hydrotheca each; non-thecate internodes rare but sometimes present; hydrothecae nearly equal in depth and breadth; margin flaring; septal ridges absent; two supracalycine nematophores, one mesial nematophore on each hydrocladial internode and one or two in the axil of the hydrocladium; all monothalamic.

Gonosome.—Gonangia taking the place of hydrocladia, the processes from the internodes supporting these as they do the hydrocladia; large, irregularly oblong, truncate distally, tapering slightly proximally.

Distribution.—Santa Barbara (Nutting); Pacific Grove, shore (Torrey); Port Townshend (Calkins); Port Renfrew; Gabriola pass, Friday Harbor (Fraser); Pacific Grove.

Plumularia halecioides Alder
Plate 43, Fig. 227

Plumularia halecioides ALDER, Ann. and Mag. N.H., (3), III, 1859, p. 353.

HINCKS, Br. Hyd. Zooph., 1868, p. 306.

FRASER, Alaskan Hyd., 1914, p. 222.

Trophosome.—Stem fascicled, about 25 mm. high; irregularly branched; branches given off from different aspects of the stem, fascicled towards the base; hydrocladia alternate, distant, each arising from a shoulder near the distal end of the internode; short, never bearing more than four hydrothecae; divided into alternate thecate and non-thecate internodes, the basal being non-thecate; the thecate internodes are much longer than the non-thecate. The hydrotheca, situated near the distal end of the internode, is nearly equal in depth and breadth, and is but slightly flaring, if at all. There is a nematophore near the proximal end of each thecate internode and two just above or beside the upper part of the hydrotheca; one on each non-thecate internode with the exception of the proximal one, and one cauline, in the hydrocladial axil. There is no sign of interseptal ridges.

Gonosome.—"Gonothecae large, ovate, ribbed transversely, with a broad, truncated top and a very short pedicel, borne on the stem, singly or in clusters" (Hincks).

Distribution.—Off Trinity islands, gulf of Alaska, 50 fathoms (Fraser); off Klashwan point, off Virago sound. 30 to 60 fathoms.

Plumularia lagenifera Allman
Plate 43, Fig. 228

Plumularia lagenifera ALLMAN, Proc. Linn. Soc., London, 1885, p. 157.

NUTTING, Am. Hyd., I, 1911, p. 65.
FRASER, West Coast Hyd., 1911, p. 82.
Vancouver Island Hyd., 1914, p.207.
A new Hydractinia, *etc.*, 1922, p. 99.
Hyd. from west coast of V.I., 1935,
p. 145.
Hyd. Distr. in vicinity of Q.C.I.,
1936, p. 126.

Trophosome.—Colonies plumose, from 5 to 10 cm. in height, growing in clusters; stem simple, divided into regular internodes, each of which bears a hydrocladium; the hydrocladia are alternate but are not in the same plane, two in succession make an angle of 100°-120° with each other; they are short, seldom branched, divided into alternate non-thecate and thecate internodes, the proximal being non-thecate; it is shorter than the intermediate internodes and has but one septal ridge, while each of the others has two. In each thecate internode there are usually three ridges well marked, one at each end and one at the base of the hydrotheca; the hydrotheca is much nearer the distal than the proximal end of the internode, its depth and breadth are nearly equal; in most cases, the internode is swollen below the hydrotheca. There are two supracalycine nematophores, a mesial one on each hydrocladial internode with exception of the first, one on each cauline internode, on the side opposite the hydrocladial process and two in the axil of the process.

Gonosome.—Gonangia borne on the hydrocladial processes of the cauline internodes; in the one diameter, the gonangium is oval, with a distinct neck, which is sometimes slightly curved, and a terminal opening, the proximal end is tapered to form a short pedicel; in the other diameter, it is much compressed.

Distribution.—Puget sound (Steindachner); coast of California (Clark); Vancouver island (Allman); Victoria, Berg inlet, Popoff island (Nutting); San Pedro, Santa Cruz, Catalina island (Torrey); San Juan archipelago, Port Renfrew, Ucluelet, Dodds narrows, Hope island; Amphitrite point, Swiftsure shoal, off Massett, Northumberland channel, Gabriola pass; in nearly all shore collections between Long beach and Esperanza inlet on the west coast of Vancouver island; entrance to Flamingo harbour, entrance to Big bay, in various locations in Houston Stewart channel, in tow net near surface off cape St. James, Massett harbour, outside Massett inlet (Fraser); San Diego, Catalina island, Santa Cruz, Long beach, San Pedro, off Pinos light, Aumentos rock, 11 locations, generally distributed, in

San Francisco bay; off Heceta head, Ore.; Nawhitti bar, western portion of Goose island halibut grounds, off Klashwan point; Sitka harbour, Alaska. Low tide to 80 fathoms.

Plumularia megalocephala Allman
Plate 43, Fig. 229

Plumularia megalocephala ALLMAN, Mem. Mus. Comp. Zool., 1877,
p. 31.
NUTTING, Am. Hyd., I, 1900, p. 57.
FRASER, West Coast Hyd., 1911, p. 83.

Trophosome.—"Hydrocaulus irregularly branched, not fascicled; pinnae alternate, each borne close to the distal end of an internode; proximal internode of pinna short and destitute of hydrothecae; following internodes longer, every alternate one carrying a hydrotheca, and slightly longer than the others. Hydrothecae small and shallow, each borne near the middle of its internode, and supporting a very large hydranth. Besides the supracalycine nematophores, each hydrotheca-bearing internode carries a single mesial nematophore at the proximal side of the hydrotheca; intervening internode carries two mesial nematophores, the short basal internode carries one" (Allman).

Gonosome.—Unknown.

Distribution.—Off San Diego in 40-75 fathoms, off Alligator reef, 14 fathoms (Allman).

Plumularia plumularoides (Clark)
Plate 44, Fig. 230

Halecium plumularoides CLARK, Alaskan Hyd., 1876, p. 217.
Plumularia plumularoides NUTTING, Am. Hyd., I, 1900, p. 62.
FRASER, West Coast Hyd., 1911, p. 84.
Monobrachium parasitum, *etc.*,
1918, p. 136.
Hyd. from west coast of V.I.,
1935, p. 145.

Trophosome.—Colonies growing together in bunches, reaching a height of 30 mm.; stem simple, divided into regular internodes by well marked nodes, each bearing a single hydrocladium on a prominent process near the distal end, the hydrocladia alternate but in the same plane. The first hydrocladial internode is short and does not bear a hydrotheca, but all the others, as many as five, are thecate, each bearing one hydrotheca; hydrotheca nearly equal in depth and breadth; septal ridges absent; two supracalycine nematophores, one mesial nematophore on each hydrocladial internode and one or two at the axil of the hydrocladium on the cauline internodal process; all monothalamic.

Gonosome.—Gonangia attached either to the process that supports the hydrocladium or to the thecate internodes, just lateral to the hydrothecae, similar in size and shape to those of *P. lagenifera*, oval, but greater in one transverse diameter than in the other, narrowing to a small process of attachment proximally and extending into a bottle neck with a small circular opening, distally. The male and the female gonangia are similar in size and shape; the male blastostyle has a pair of processes near the centre, projecting outward and slightly backward.

Distribution.—Cape Etolin, 8 to 10 fathoms, Nunivak island, Alaska (Clark); San Diego, 15-25 fathoms (Torrey); on *Macrocystis* at the entrance to Bull harbour, Hope island; Catala island, west coast of Vancouver island, low tide (Fraser).

Plumularia setacea (Ellis)

Plate 44, Fig. 231

Corallina setacea ELLIS, Nat. Hist. Corallines, 1755, p. 19.
Plumularia setacea NUTTING, Am. Hyd., I, 1900, p. 56.
                    FRASER, West Coast Hyd., 1911, p. 84.
                            Vancouver Island Hyd., 1914, p. 209.
                            Hyd. from west coast of V.I., 1935, p. 145.
                            Hyd. Distr. in vicinity of Q.C.I., 1936, p. 126.

Trophosome.—Colony not large, sometimes reaching a height of 50 mm., but often much less; stem simple, divided into regular internodes, each bearing a hydrocladium, which is seldom branched; the hydrocladia are regularly alternate and are in the same plane. After the first internode, which is short, and without a hydrotheca, thecate and non-thecate internodes alternate; the hydrotheca is placed near the distal end of the internode; in many cases, scarcely any internodal septa can be seen, but in other cases, there may be an indication of as many as there are in *P. lagenifera*, in some parts of the colony; there are two supracalycine nematophores, one mesial nematophore to each hydrocladial internode, with the exception of the proximal, one on each cauline internode on the side opposite the hydrocladial process and one in the axil of that process.

Gonosome.—Gonangia borne on the hydrocladial processes of the cauline internodes, much elongated, usually with a long neck and a circular aperture.

Distribution.—Santa Barbara (Nutting); point Wilson (Calkins); San Diego, La Jolla, Avalon, Catalina island, San Pedro, point Loma, Monterey, San Francisco, Cal., 1-25 fathoms; Victoria (Torrey); Pacific Grove (Stechow); San Juan archipelago; Ucluelet, Northumberland channel, Porlier pass, Friday Harbor; off Bajo reef, bar off Indian Village in Esperanza inlet; in tow net near surface, western portion of Houston Stewart channel (Fraser); San Diego, Coronado island, La Jolla, point Loma, point Pinos, Wilson's cove, San Clemente, Avalon, Santa Monica, Monterey; outside of Golden Gate and several locations in the middle and lower sections of San Francisco bay; Heceta head, Ore.; 16 miles northeast of Reef island, Q.C.I.; Sitka, Alaska.   Low tide to 90 fathoms.

Plumularia virginiae Nutting
Plate 44, Fig. 232

Plumularia virginiae NUTTING, Am. Hyd., I, 1900, p. 66.
FRASER, West Coast Hyd., 1911, p. 85.

Trophosome.—"Colonies growing in tufts of simple plumose stems, attaining a height of about half an inch; stem not fascicled, divided into regular internodes, each of which bears a hydrocladium, on a short process from near its distal end and shows a thickened internal ridge near each end; hydrocladia alternate, not very closely approximate.   Proximal internode short, with a single internal ridge; all of the remaining internodes are hydrothecate, each with a very strong internal thickening on its anterior side just below the mesial nematophore, and another below the supracalycine pair. Hydrothecae borne just above the middle of the internodes on very strong shoulders or protuberances, very shallow, basin-shaped, with broadly flaring sides; hydranths very large, robust, with about 24 tentacles and a broadly expanded hypostome, reminding one of *P. halecioides*.   Nematophores long, conical, with a very shallow distal chamber, and containing sarcostyles with remarkably symmetrical batteries of small nematocysts, and the usual sarcodal process; supracalycine nematophores borne on prominent swellings of the internodes, and directed upward and outward; a mesial nematophore near the proximal end of each internode and a cauline nematophore in the axil of each hydrocladium" (Nutting).

Gonosome.—"Gonangia borne in a row on the front of the stem, long, with the distal end produced into a neck, as in *P. setacea*. The younger gonangia are long, conical bodies with truncated distal ends" (Nutting).

Distribution.—Santa Barbara, Cal. (Nutting).

Genus TETRANEMA new genus

Trophosome.—Stem fascicled; hydrothecal margin entire; one mesial nematophore and two pairs of lateral nematophores to each hydrotheca; no supracalycine nematophores.

Gonosome.—Gonangia borne on the hydrocladia without protective structures.

Tetranema furcata new species
Plate 44, Fig. 233

Trophosome.—Colony 35 mm. high; stem fascicled in the proximal portion but rather slender; the hydrocladia are all borne on the one tube, which is definitely divided into nodes but the other tubes are not; there are broad, low, sessile nematophores on all tubes. Hydrocladia are given off alternately, one from each internode, from a prominent process near the middle of the internode; there is a hydrotheca in each axil, smaller, more definitely tubular and with thicker wall than the regular hydrocladial hydrotheca; some of the distal hydrocladia are branched, or more definitely speaking, are divided into two equal, or nearly equal, branches, with angle between them, acute. The hydrocladium is divided into internodes by distinct nodes; the first internode is short, without a hydrotheca, but with a mesial nematophore; each of the other internodes is long and slender with the hydrotheca less than half the length of the internode, somewhat nearer the distal end; hydrothecae more than twice as deep as wide, almost tubular but somewhat inflated in the proximal portion, adherent to the hydrocladium throughout; margin entire but slightly sinuous; there are two pairs of low, sessile, lateral nematophores and a mesial nematophore proximal to the hydrotheca. There are no supracalycine nematophores. Hydrocladial and intrathecal septa absent.

Gonosome.—Gonangia growing singly from the hydrocladium, between the mesial nematophore and the hydrotheca, 1.5 mm. long, elongate-obovate, with the distal end sharply curved so that the opening almost faces the base; no definite pedicel.

Distribution.—Kaison bank, west coast of Moresby island, Q.C.I. 110 fathoms.

LITERATURE CITED

AGASSIZ, A.

  1865. North American Acalephae. Ill. Cat. Mus. Comp. Zool. at Harvard
    College, 2: 1-234.

AGASSIZ, L.

  1862. Contributions to the Natural History of the United States of America,
    4: 1-372.

ALDER, J.

  1856. A notice of some new genera and species of British Hydroid Zoophytes.
    Ann. and Mag. Nat. Hist. (2), 18: 353-362. London.
  1856. Description of three new British Zoophytes. *Ibid.*, 353-362.
  1857. A catalogue of the Zoophytes of Northumberland and Durham.
    Trans. Tyneside Nat. Field Club, 3: 1-70. Newcastle-upon-Tyne.
  1859. Description of three new species of sertularian Zoophytes. Ann. and
    Mag. Nat. Hist. (3), 3: 353-355.
  1862. Description of some rare Zoophytes found on the coast of Northum-
    berland. *Ibid.*, (3), 9: 311-316.
  1864. Supplement to the Catalogue of the Zoophytes found on the coast
    of Northumberland and Durham. Trans. Tyneside Nat. Field Club,
    5: 1-23.

ALLMAN, G. J.

  1859. Notes on the hydroid polyps. Ann. and Mag. Nat. Hist. (3),
    4: 137-144.
  1863. Notes on the Hydroida. *Ibid.*, (3), 11: 1-12.
  1864. On the construction and limitation of genera among the hydroids.
    Ibid., (3), 13: 345-380.
  1864. Notes on the Hydroida. *Ibid.*, (3), 14: 57-64.
  1871. A monograph of the gymnoblastic or tubularian hydroids. Published
    for the Ray Society, in 2 parts, 450 pp. London.
  1877. Report on the Hydroida collected during the exploration of the
    Gulf Stream by L. F. de Pourtales. Mem. Mus. Comp. Zool. at
    Harvard College, 5, No. 2; 1-64.
  1883. Report on the Hydroida dredged by H.M.S. Challenger, during the
    years 1873-1876, pt. 1, Plumularidae. Voyage of the Challenger,
    20: 1-54. London.
  1885. New Hydroida from the collection of Miss H. Gatty. Jour. of the
    Linn. Soc., 19: 132-161. London.
  1888. Report on the Hydroids dredged by H.M.S. Challenger, during the
    years 1873-1876, pt. 2, Voyage of the Challenger, 23: 1-87. London.

BALE, W. M.

  1888. On some new and rare Hydroida in the Australian Museum Coll.
    Proc. Linn. Soc., N.S.W., (2) 3: 745-799.
  1893. Further notes on Australian Hydroids with description of some new
    species. Proc. Royal Soc. Victoria, 93-117.

BENEDEN, P. J. VAN
   1844. Mémoire sur les Campanulaires de la Cote d'Ostende, considérés
   sous le rapport physiologique, embryogénique et zoologique. Nouv.
   Mém. de l'Academie Royal, Bruxelles, 18: 1-42.

BERGH, R. S.
   1887. Goplepolyper (Hydroider) fra Kara Havet. Dijmphna zoologisk-
   botaniske udbytte. 333-337. Kjobenhavn.

BIGELOW, H. B.
   1913. Medusae and Siphonophorae collected by the U.S. Fisheries Steamer
   "Albatross" in the northwestern Pacific, 1906. Proc. U.S. Nat. Mus.,
   44: 1-119.

BILLARD, A.
   1907. Hydroides. In: Expeditiones Scientifiques du "Travailleur" et du
   "Talisman", 8: 139-241. Paris.

BONNEVIE, K.
   1898. Neue Norwegischen Hydroiden. Bergens Museum Aarbog, 5: 1-15.
   1899. Den norske Nordhavsexpedition, 1876-78, 6: pt. 26: Zoologi Hydroida,
   1-103.

BROCH, H.
   1909. Die Hydroiden der Arktischen Meere. Fauna Arctica, 5: 129-248.
   Jena.
   1909. Hydroidunterzuchungen II. Zur Kenntnis der Gattungen Bonneviella and Lictorella. Nyt Magazin for Natursvidenskaberne, 47:
   195-205. Christiania.
   1912. Coelenterés du Fond. Duc d'Orleans Campagne Arctique de 1907.
   1-28. Bruxelles.

BROWNE, E. T.
   1907. The Hydroids collected by the "Huxley" from the north side of the
   bay of Biscay in August 1906. Jour. Marine Biol. Assn., 8, No. 1:
   13-36.

CALKINS, G. N.
   1899. Some hydroids from Puget sound. Proc. Boston Soc. Nat. Hist.,
   28: 333-367.

CLARK, S. F.
   1876. Description of new and rare hydroids from the New England coast.
   Trans. Conn. Acad. Sc., 3: 58-66. New Haven.
   1876. The hydroids of the Pacific coast of the United States south of Vancouver island, with a report upon those in the Museum at Yale College.
   Ibid., 3: 249-264.
   1876. Report of the hydroids on the coast of Alaska and the Aleutian
   islands collected by W. H. Dall, from 1871 to 1874. Proc. Acad. Nat.
   Sc., Phila., 205-238. Philadelphia.

CLARKE, S. F.
    1894. Report on the dredging operations off the west coast of Central America to the Galapagos, to the west coast of Mexico, and in the gulf of California, in charge of Alexander Agassiz, carried on by the U.S. Fish Commission Steamer "Albatross" during 1891, Lieut. Commander Z. I. Tanner, U.S.N., commanding. The Hydroids. Bull. Mus. Comp. Zool. at Harvard College, 25, no. 6: 71-77.

ELLIS, J.
    1755. An essay towards the natural history of the corallines and other marine productions of the like kind found off the coasts of Great Britain and Ireland. London.

ELLIS, J. and D. SOLANDER
    1786. The natural history of many curious and uncommon zoophytes collected from various parts of the globe. 208 pp. London.

FABRICIUS, O.
    1780. Fauna Groenlandica. Hauniae et Lipsiae.

FEWKES, J. W.
    1889. New Invertebrata from the coast of California. 1-50.
    1891. An aid to a collector of the Coelenterata and Echinodermata of New England. Bull. Essex Inst., 23: 1-90.

FLEMING, J.
    1828. A history of British Animals. Edinburgh.

FORBES, E.
    1848. A monograph of the British naked-eye medusae. Ray Society. London.

FRASER, C. McL.
    1911. The hydroids of the west coast of North America. Bull. from the Laboratories of Nat. Hist., State Univ. of Iowa, 1-91. Iowa City.
    1912. Notes on New England Hydroids. Ibid., 39-48.
    1912. Some Hydroids of Beaufort, North Carolina. Bull. Bureau of Fisheries, 30: 339-387. Washington.
    1913. Hydroids from Vancouver island. Can. Geol. Surv., Victoria Mem. Mus., Bull. 1, pt. 15: 147-155. Ottawa.
    1913. Hydroids from Nova Scotia. Ibid., pt. 16: 157-186.
    1914. Some hydroids of the Vancouver island region. Trans. Royal Soc. Can., (3), 8, Sect. 4: 99-216. Ottawa.
    1914. Notes on some Alaskan hydroids. Ibid., 217-222.
    1918. Monobrachium parasitum and other west coast hydroids. Ibid., (3), 12, Sect. 4: 131-138.
    1918. Hydroids of eastern Canada. Contr. to Can. Biol., 329-367. Ottawa.
    1921. Canadian Atlantic Fauna, 3a Hydroida, 1-46. Ottawa.
    1922. A new Hydractinia and other west coast hydroids. Contr. to Can. Biol., 97-100.

1925. Some new and some previously unreported hydroids, mainly from the California coast. Univ. Calif. Pub. Zool., 28, no. 7: 167-172. Berkeley.

1926. Hydroids of the Miramichi estuary collected in 1918. Trans. Royal Soc. Can., (3), 20, sect. 5: 209-214.

1933. Hydroids as a food supply. *Ibid.*, (3), 27: 259-264.

1935. Hydroids from the west coast of Vancouver island. Can. Field-Naturalist, 44, no. 9: 143-145. Ottawa.

1936. Hydroids from the Queen Charlotte islands, Jour. Biol. Board Can., 1, pt. 6: 503-507.

1936. Hydroid distribution in the vicinity of the Queen Charlotte islands. Can. Field-Naturalist, 50, no. 7: 122-126.

HARTLAUB, C.

1897. Die Hydromedusen Helgolands. Wissen. Meeresuntersuchungen, n.f., 2, heft 1: 449-514. Kiel und Leipzig.

1901. Hydroiden aus dem Stillen Ocean. Ergebnisse einer Reise nach dem Pacific, 1896-97. Zool. Jahrb., 14, hft. 5: 349-379. Jena.

1905. Die Hydroiden der magalhaenischen Region und chilienschen Küste. Fauna Chilensis, 3, hft. 3: 497-718. Jena.

HASSALL, A.

1852. Description of three species of marine zoophytes. Trans. Royal Micr. Soc., 3. London.

HINCKS, T.

1853. Further notes on British zoophytes, with descriptions of new species. Ann. and Mag. Nat. Hist., (2), 11: 178-185.

1861. A catalogue of the zoophytes of South Devon and South Cornwall. *Ibid.*, (3), 8: 152-161, and pages 353-362.

1863. On some new British hydroids. *Ibid.*, (3), 11: 45-47.

1866. On new British hydroids. *Ibid.*, (3), 18: 296-299.

1868. A history of the British Hydroid Zoophytes. 2 vols. London.

1874. On deep water hydroids from Iceland. Ann. and Mag. Nat. Hist., (4), 13: 146-153.

JÄDERHOLM, E.

1903. Aussereuropaische Hydroiden in swedischen Reichmuseum. Arkiv for Zool., 1: 259-312. Stockholm.

1907. Zur Kenntnis der Hydroidenfauna des Beringsmeeres. *Ibid.*, 4: no. 8: 1-8.

1909. Northern and Arctic invertebrates in the collection of the Swedish State Museum. IV. Hydroiden. Kongelige Svenska Vetenskaps Akademiens Handlingar, 45, no. 1: 1-124. Stockholm.

JOHNSTON, G. H.

1847. History of British Zoophytes. Ed. II. 2 vols. London.

KIRCHENPAUER, G. H.
  1872. Ueber die Hydroidenfamilie Plumularidae, einzelne gruppen derselben
         und ihre Fruchtbehalter. I. Aglaophenia. Abhandlungen aus dem
         Gebiete der Naturwissenschaften herausgeben von dem Naturwis-
         senschaftlichen Verein in Hamburg. 1-52.
  1884. Nordische Gattungen und Arten von Sertulariden. *Ibid.*, 8, pt. 3:
         1-54.

KRAMP, P.
  1911. Report on the hydroids collected by the Denmark Expedition at
         Northeast Greenland. Danmark Expedition til Groenlands Nordoest-
         kyst, 1906-1908, bind V.M. 7.   Saertryk of "Meddelelser om Groen-
         land", 45: 341-396. Copenhagen.
  1932. The Godthaab Expedition 1928 Hydroids. *Ibid.*, 79, no. 1: 1-86.

LAMARCK, J. B. P. VON
  1836. Histoire Naturelle des Animaux sans Vertèbres. 2nd. Ed.

LEPECHIN, J.
  1781. Novae Pennatulae et Sertulariae species descriptae. Acat Acad.
         Sc. Imp. Petropolitana pro anno 1778. *Fide* Broch.
  1783. Sertulariae species duae determinatae. *Ibid.*, 1780.

LESSON, R. P.
  1843. Hist. Nat. des Zoophytes Acalèphes. Paris.

LEVINSEN, G. M. R.
  1892. Om Fornyelsen Ernaeringsindividerne hos Hydroiderne. Vidensk.
         Meddel. fra den naturh. Foren. Kjobenhavn. 12-31.
  1893. Meduser, Ctenophorer og Hydroider fra Groenlands Vestkyst
         tilligemed Bemaerkninger on Hydroidernes Systematik. *Ibid.*, 143-220.

LINNAEUS, C.
  1758. Systema naturae, 10th ed. Lipsiae.
  1767. Systema naturae, 12th ed. Holmiae.

McCRADY, J.
  1858. Gymnophthalmata of Charleston harbor. Proc. Elliott Soc. Nat.
         Hist., 1 (for 1853-1858): 103-221. Charleston.

MacGILLIVRAY, J.
  1842. Catalogue of the marine zoophytes of the neighborhood of Aberdeen.
         Ann. and Mag. Nat. Hist., (1), 9: 462-469.

MARKTANNER-TURNERETSCHER, G.
  1890. Die Hydroiden des K.K. naturhistorischen Hofsmuseums. Ann.
         des K.K. naturh. Hofsmuseums, 5: 195-286. Wien.
  1895. Hydroiden in Zoolog. Ergebnisse im Jahre 1889 von Dr. W. Küken-
         thal und Dr. A. Walter ausgeführten Exped. nach Ost-Spitzbergen.
         Zool. Jahrb. Abt. fur Syst. 8: 391-438.

MAYER, A. G.

1910. Medusae of the world. 3 vols. Carnegie Inst. Washington.

MERESCHKOWSKY, M. C.

1877. On a new genus of hydroids from the White Sea, with a short description of other new hydroids. Ann. and Mag. Nat. Hist., (4), 20: 220-229.
1878. New Hydroida from Ochkotsk, Kamstschatka, and other parts of
the North Pacific Ocean. Ibid., (3), 2: 433-451.

MURRAY, A.

1860. Description of new Sertularidae from the California coast. Ibid.,
(3), 5: 250-252 and 504.

NORMAN, A. M.

1864. On undescribed British Hydroida, Actinozoa and Polyzoa. Ibid.,
(3), 13: 82-84.

NUTTING, C. C.

1899. Hydroida from Alaska and Puget sound. Proc. U.S. Nat. Mus.,
21: 741-753.
1900. American Hydroids. Pt. I. The Plumularidae. Special Bulletin,
U.S. Nat. Mus., 142 pp.
1901. The hydroids of the Woods Hole region. U.S. Fish Comm. Bull.
for 1899, 19: 325-386.
1901. Papers from the Harriman Alaska Expedition. XXI—The Hydroids.
Proc. of the Washington Acad. Sc., 3: 157-216.
1904. American Hydroids. Pt. II. The Sertularidae. Special Bull.
U.S. Nat. Mus., 152 pp.
1905. Hydroids of the Hawaiian islands collected by the Steamer "Albatross" in 1902. Bull. U.S. Fish Comm. for 1903, 931-959.
1915. American Hydroids. Pt. III. The Campanularidae and the Bonneviellidae. Special Bull. U.S. Nat. Mus. 126 pp.
1927. Report on the Hydroida collected by the United States Fisheries
Steamer "Albatross" in the Philippine region, 1907-1910. U.S. Nat.
Mus. Bull. 100, 6, pt. 3: 195-242.

PALLAS, P. S.

1766. Elenchus Zoophytorum.

PERON et LESUEUR

1809. Tableau des Caractères Génériques et Spécifiques de toutes les
Espèces de Meduses commes jusqu'a ce jour. Annales du Muséum
d'Histoire Naturelle, 14. Fide Forbes.

PICTET, C. et M. BEDOT

1900. Hydraires provenant des Campagne de l'Hirondelle (1886-1888).
Fasc. xviii, 1-55.

RITCHIE, J.
1907. The Hydroids of the Scottish National Antarctic Expedition. Trans Royal Soc. Edin., 45, pt. 2, no. 18: 519-545.
1909. Supplementary Report on the hydroids of the Scottish National Antarctic Expedition. *Ibid.*, 47, pt. 1, no. 4: 65-101.
1912. Some northern hydroid zoophytes obtained by Hull trawlers; with a description of a new species of Plumularian. Proc. Royal Phys. Soc. Edin., 18, no. 4: 219-230.

SARS, G. O.
1873. Bidrag til Kundskaben om Norges Hydroider. Videns.-Selsk. Forh. for 1872. Christiania.

SARS, M.
1851. Beretning om en i Sommeren 1849 foretagen Zoologisk Reise i Lofoten of Finmarken. Nyt Magazin for Naturvidenskaberne, 6. Christiania.
1857. Bidrag til Kundskaben om Middlehavets Littoralfauna. *Ibid.*, 10.
1863. Bemerkninger over fire norske Hydroider. Vidensk.-Selsk. Forh. for 1862.

SCHNEIDER, K. C.
1897. Hydropolyper von Rovigno nebst uebersicht über das System der Hydropolypen im Allgemein. Zool. Jahrb., 10: 472-555.

STECHOW, E.
1909. Hydroidpolypen der Japanischen Ostküste. Beitrage zur Natur-geschichte Ostasiens. I Teil. Athecata und Plumularidae. 1-111.
1921. Über Hydroiden der Deutschen Tiefsee-Expedition, nebst Bemer-kungen einige andre Formen. Zool. Anz., 53, nos. 9 and 10: 223-236.
1923. Neue Hydroiden der Deutschen Tiefsee-Expedition, nebst Be-merkungen über einige andere Formen. Zool. Anz., 56: 1-20.
1923. Zur Kenntnis der Hydroiden Fauna des Mittelmeeres, Amerikas und anderer Gebiete. II Teil. Zool. Jahrb. 47: 29-270.

STIMPSON, W.
1854. Synopsis of the marine invertebrata of Grand Manan. Smithsonian Contr. to Knowledge. 6: Acalephae, 8-11.

THORNELY, L. R.
1904. On the Hydroida. Report to the Government of Ceylon on the Pearl Oyster Fisheries of the gulf of Manaar. Royal Society. Supple-mentary Report 8: 107-126. London.
1908. Reports on the marine biology of the Sudanese Red Sea. X. Hydro-ida. Collected by Mr. C. Crossland from October 1904 to May 1905. Jour. Linn. Soc. in Zool., 31: 80-85.

TORREY, H. B.
1902. The Hydroida of the Pacific Coast of North America. Univ. Calif. Pub. in Zool., 1: 1-104.
1904. The hydroids of the San Diego region. *Ibid.*, 2: 1-43.

TRASK, J. B.
  1857. On some new microscopic organisms. Proc. Calif. Acad. Nat. Sc.,
         1: 110-112.

VERRILL, A. E.
  1872. Radiata from the coast of North Carolina. Am. Jour. Sci. and Arts.
         (3), 3.
  1872. Report of the invertebrate animals of Vineyard sound and adjacent
         waters. Report of the Commissioner of Fisheries for 1871-1872:
         295-747.

WARREN, E.
  1908. On a collection of hydroids mostly from the Natal coast. Annals
         of the Natal Government Museum, 1, pt. 3: 269-355.

WRIGHT, T. S.
  1857. Observations on British Zoophytes. Proc. Royal Phys. Soc. Edin.,
         1: 226-237.
  1858. Observations on British Zoophytes. *Ibid.*, 1: 447-455.
  1859. Observations on British Zoophytes. Edin. New Phil. Jour. N.S.
         8: 1-9.
  1862. On the reproduction of Thaumantias inconspicua. Quart. Jour.
         Micro. Sc., N.S., 2: 221-222 and 308.

INDEX

Synonyms are in *Italics*

PLATES

PLATE I

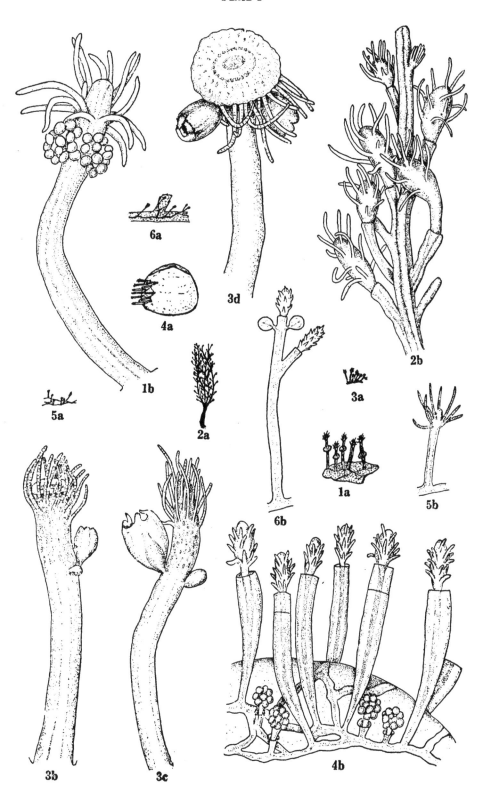

PLATE II

Fig. 7. *Monobrachium parasitum*
    a. Natural size.
    b. Colony on a bivalve shell.
    c. A zooid further enlarged (×40).
    d. Medusa buds.

Fig. 8. *Coryne brachiata*
Colony with hydrophores and gonophores (After Nutting).

Fig. 9. *Coryne corrugata*
    a. Natural size.
    b. Portion of colony with hydranths and gonophores.

Fig. 10. *Coryne crassa*
    a. Natural size.
    b. Zooid with male sporosacs.

Plate II

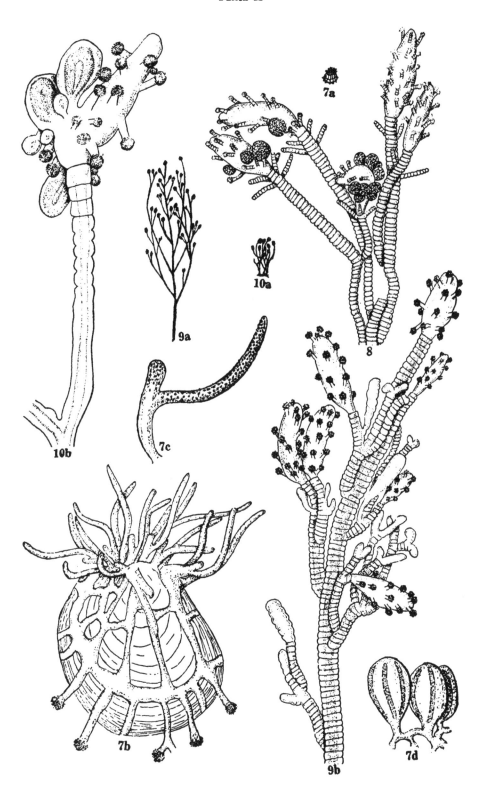

PLATE III

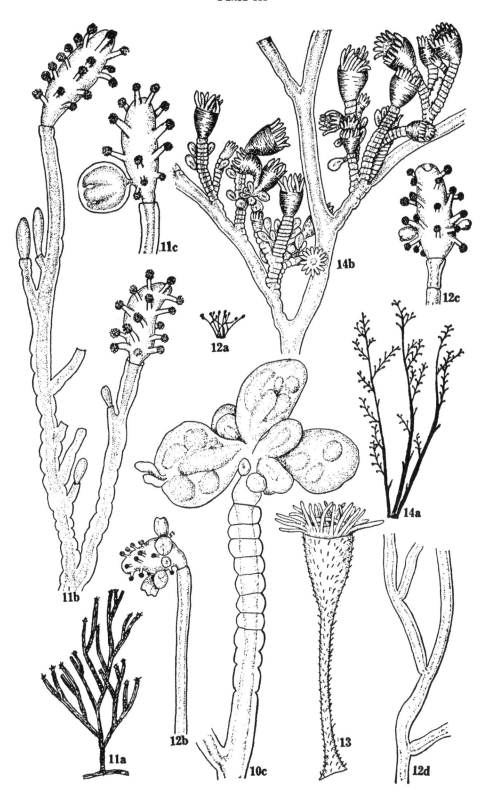

PLATE IV

Fig. 15. *Bimeria gracilis*
a. Natural size.
b. Portion of fascicled stem.
c. Portion of branch.
d. Portion of branch showing gonophores.

Fig. 16. *Bimeria pusilla*
a. Natural size.
b. A colony.

Fig. 17. *Bimeria robusta*
a. Natural size.
b. Portion of colony.

Fig. 18. *Bimeria tenella*
a. Natural size.
b. A colony.

Fig. 19. *Garveia annulata*
a. Natural size.
b. Portion of fascicled stem.

PLATE IV

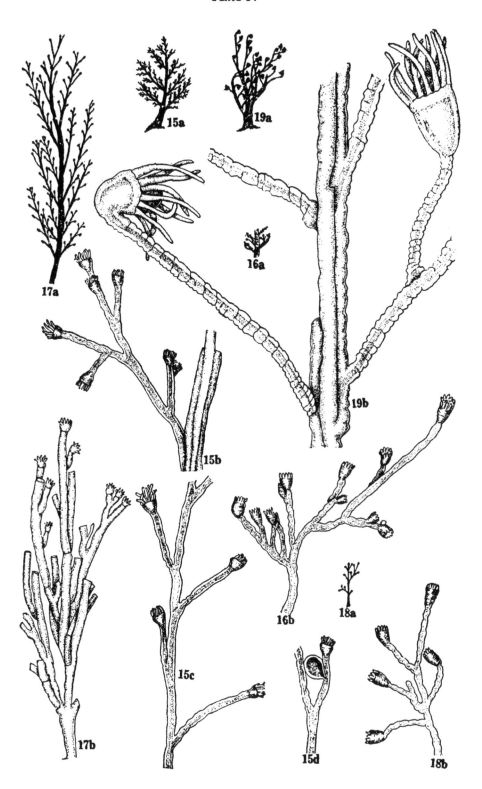

PLATE V

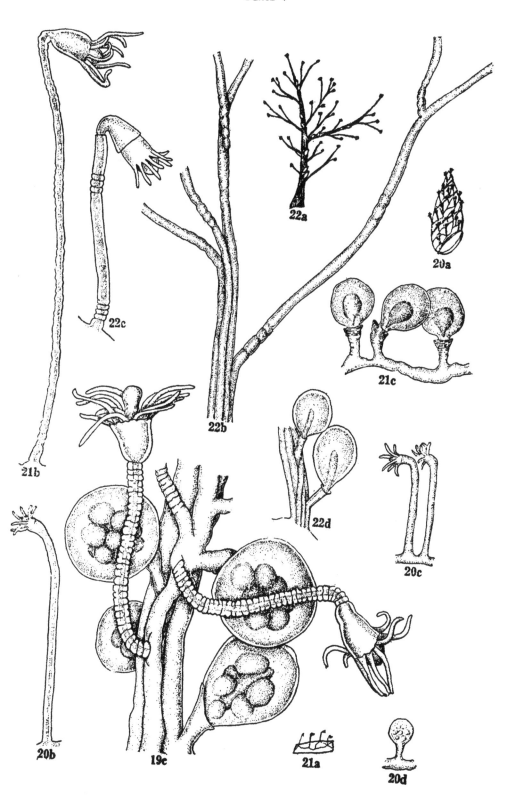

PLATE VI

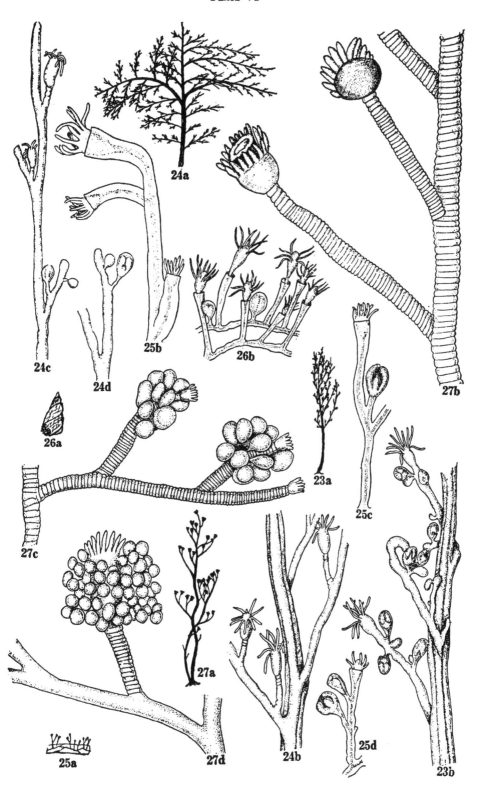

PLATE VII

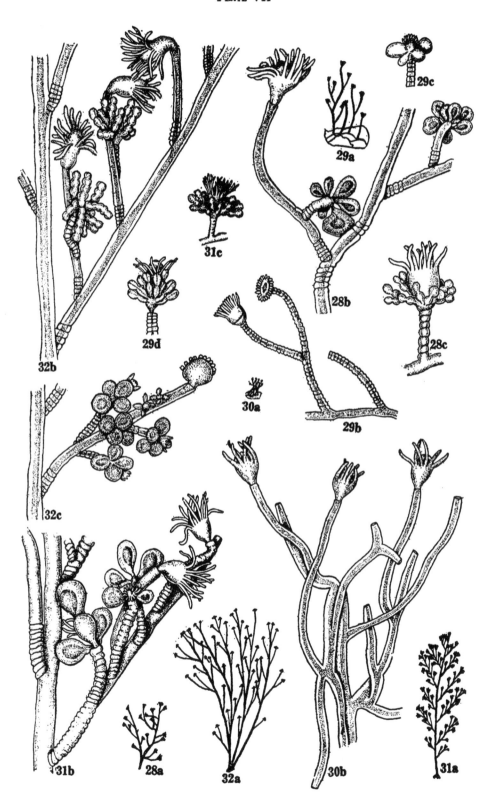

Plate VIII

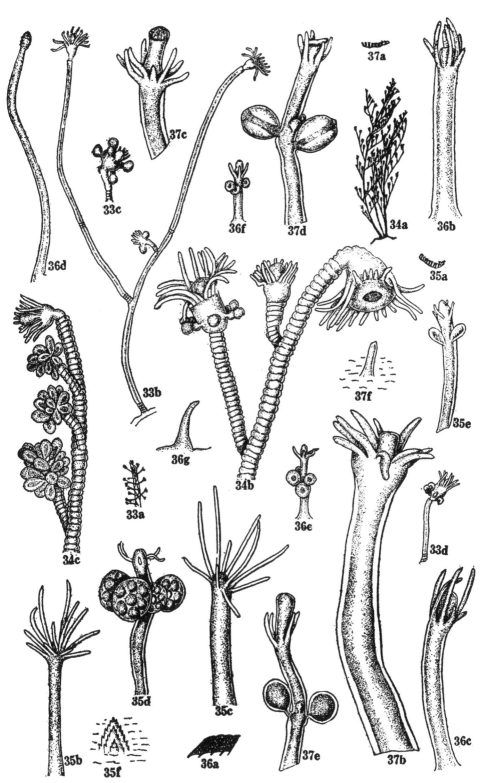

PLATE IX

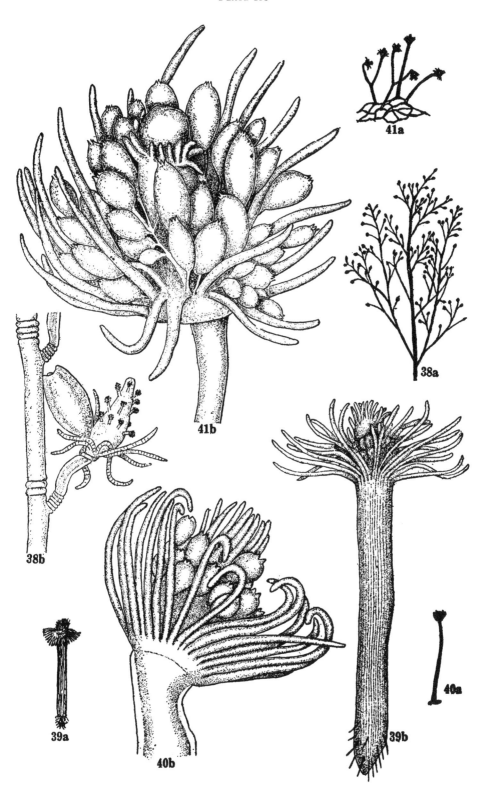

41a

38a

41b

38b

39a

40b

40a

39b

PLATE X

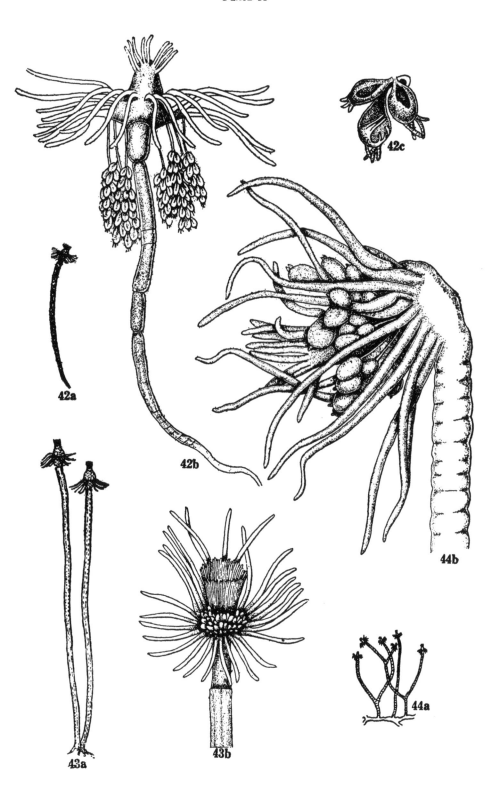

42a

42b

42c

43a

43b

44a

44b

PLATE XI

Fig. 45. *Tubularia marina*
    a. Natural size.
    b. Zooid showing gonophores.

Fig. 46. *Hybocodon prolifer*
    a. Natural size.
    b. Hydranth and gonophores (After L. Agassiz).
    c. Medusa.

PLATE XI

45a

45b

46a

46b

46c

PLATE XII

PLATE XII

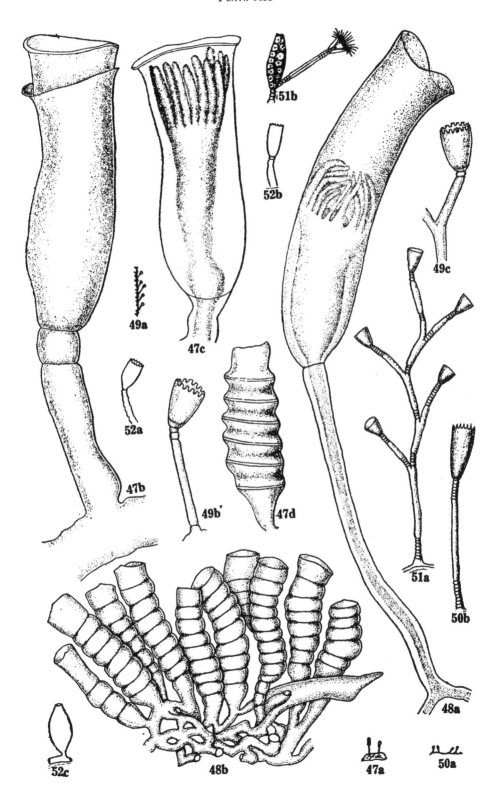

PLATE XIII

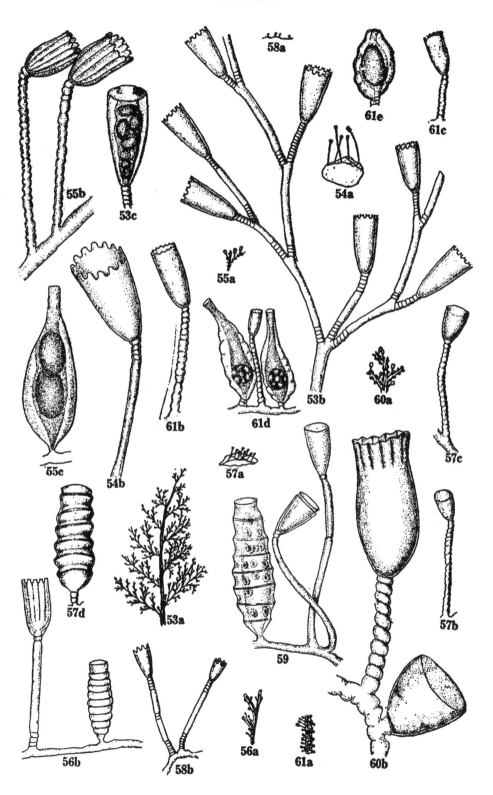

PLATE XIV

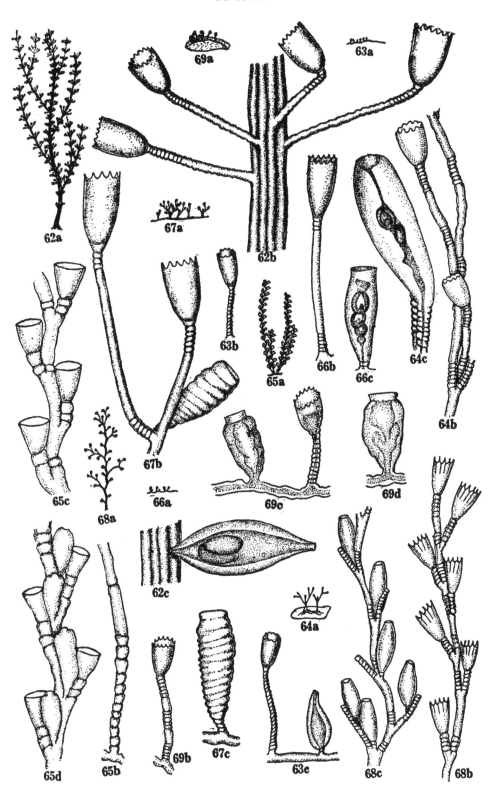

PLATE XV

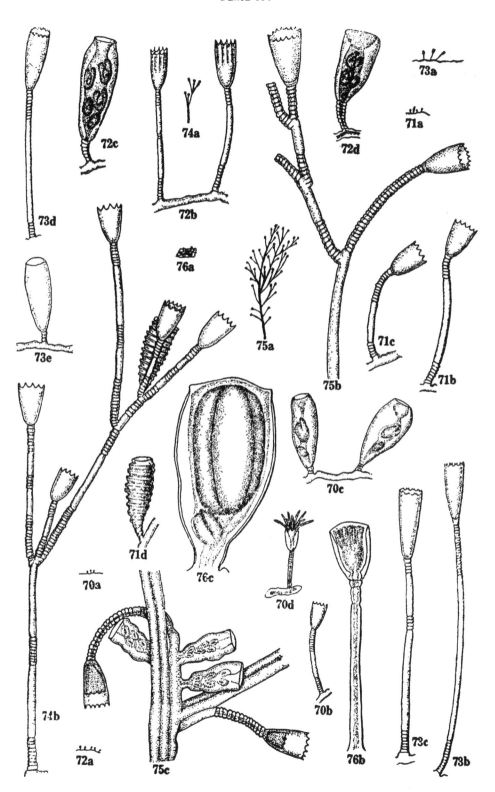

PLATE XVI

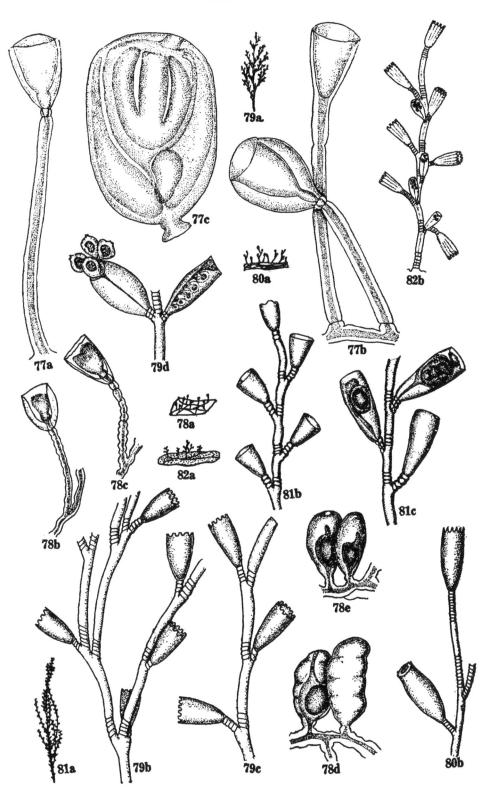

79a

77c

80a

82b

79d

77b

77a

78a

82a

81b

81c

78c

78b

78e

81a

79b

79c

78d

80b

PLATE XVII

PLATE XVIII

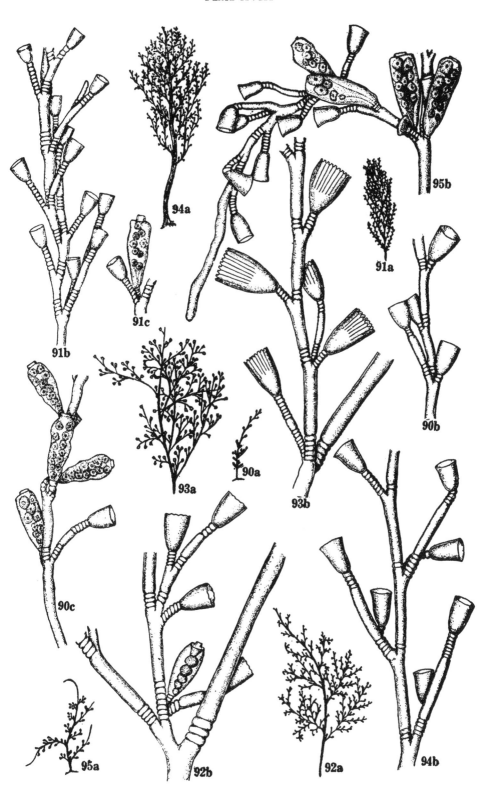

PLATE XIX

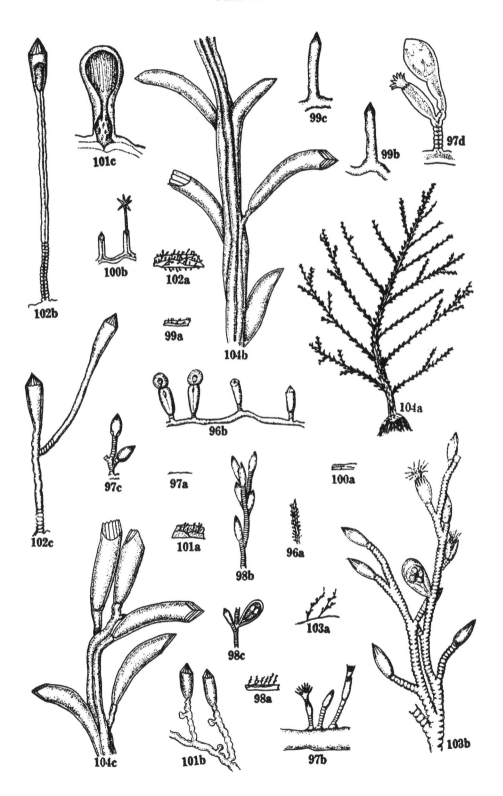

PLATE XX

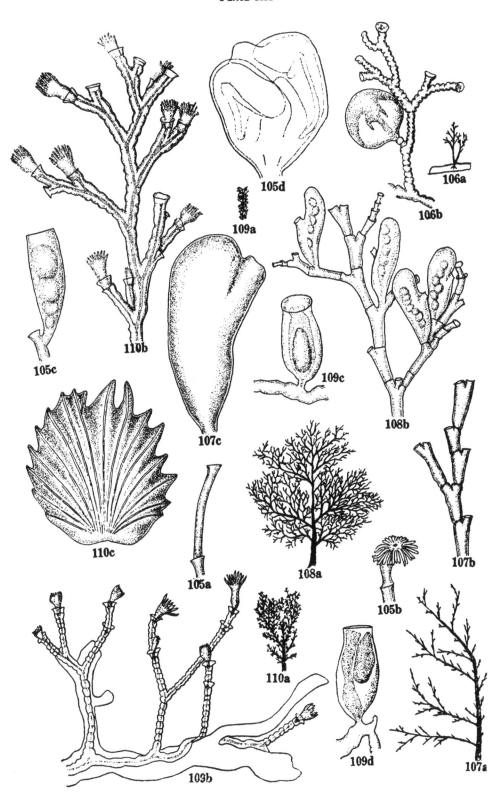

105d

109a

106a

106b

108b

110b

105c

107c

109c

110c

105a

108a

107b

110a

105b

103b

109d

107a

PLATE XXI

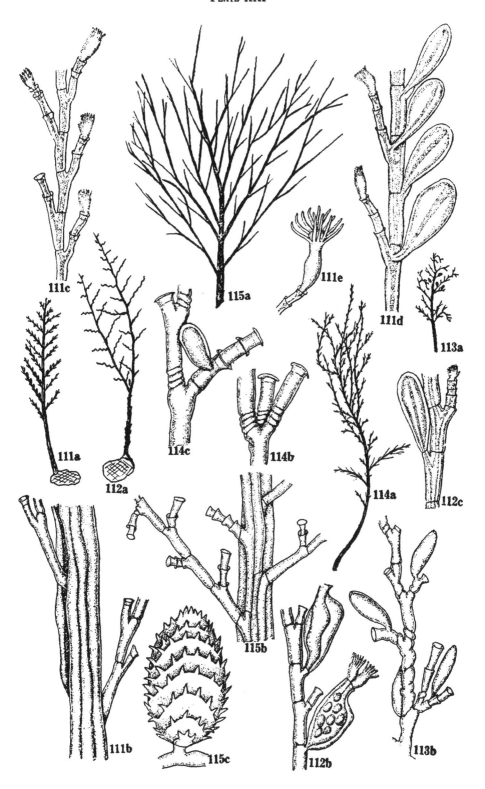

PLATE XXII

Fig. 116. *Halecium ornatum*
   a. Portion of colony, hydrophores (After Nutting).
   b. Immature gonophore (After Nutting).

Fig. 117. *Halecium parvulum*
   a. Natural size.
   b. Portion of branch showing mode of branching and male gonophores.
   c. Female gonophores.

Fig. 118. *Halecium pygmaeum*
   a. Natural size.
   b. Colony showing branching.
   c. Female gonophores.
   d. Male gonophores.

Fig. 119. *Halecium reversum*
   a. Natural size.
   b. Portion of branch showing position of nodes.
   c. Branch arising from fascicled stem

Fig. 120. *Halecium robustum*
   a. Natural size.
   b. Portion of colony.

Fig. 121. *Halecium scutum*
   a. Natural size.
   b. Portion of branch showing length of internodes.
   c. Gonophore.

Fig. 122. *Halecium speciosum*
   a. Natural size.
   b. Colony—hydrophores and gonangia (After Nutting).

PLATE XXII

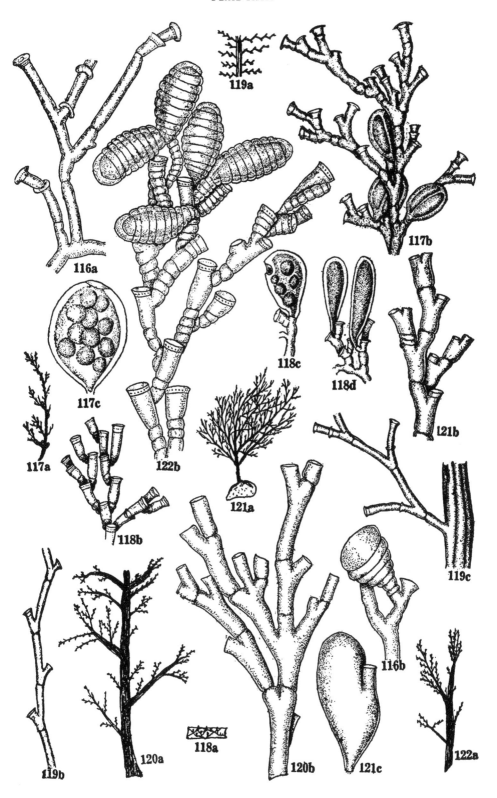

116a

119a

117b

117c

118c

118d

121b

117a

118b

122b

121a

119c

116b

118a

119b

120a

120b

121c

122a

PLATE XXIII

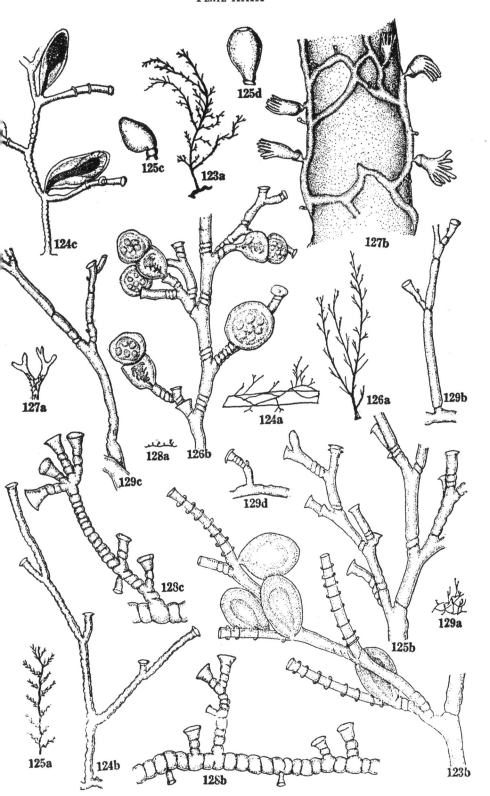

PLATE XXIV

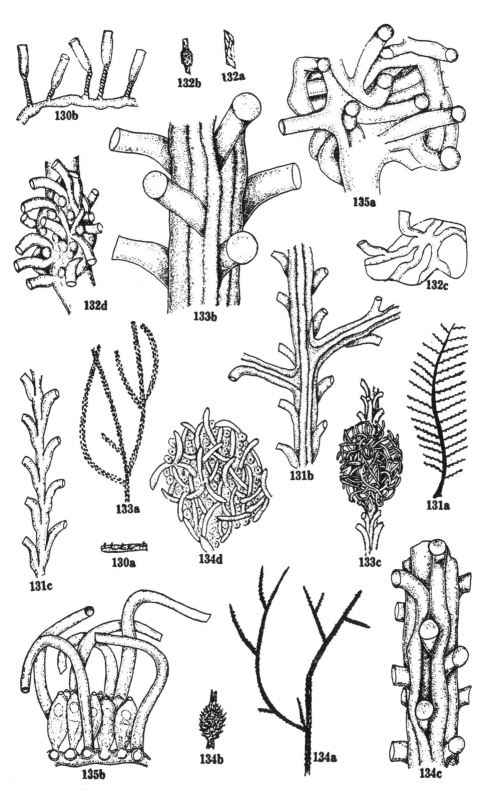

PLATE XXV

Fig. 136. *Lafoea adnata*
    a. Natural size.
    b. Portion of colony.
    c. Hydrothecae on stolon.

Fig. 137. *Lafoea dumosa*
    a. Natural size.
    b. Coppinia, natural size.
    c. Portion of branch of erect stem.
    d. Portion of creeping stem.
    e. Portion of coppinia.

Fig. 138. *Lafoea fruticosa*
    a. Natural size.
    b. Portion of branch showing hydrothecae.
    c. Portion of fascicled stem.
    d. Coppinia (After Bonnevie).

Fig. 139. *Lafoea gracillima*
    a. Natural size.
    b. Coppinia, natural size.
    c and d. Portions of stem showing hydrothecae of varying size.

PLATE XXV

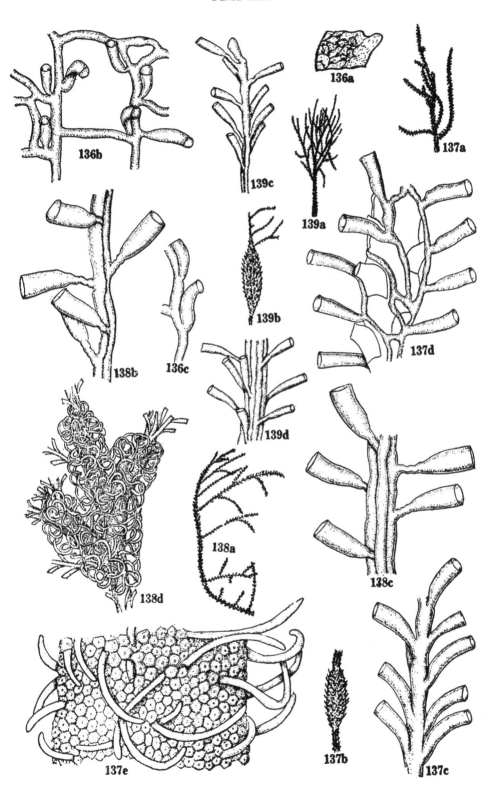

PLATE XXVI

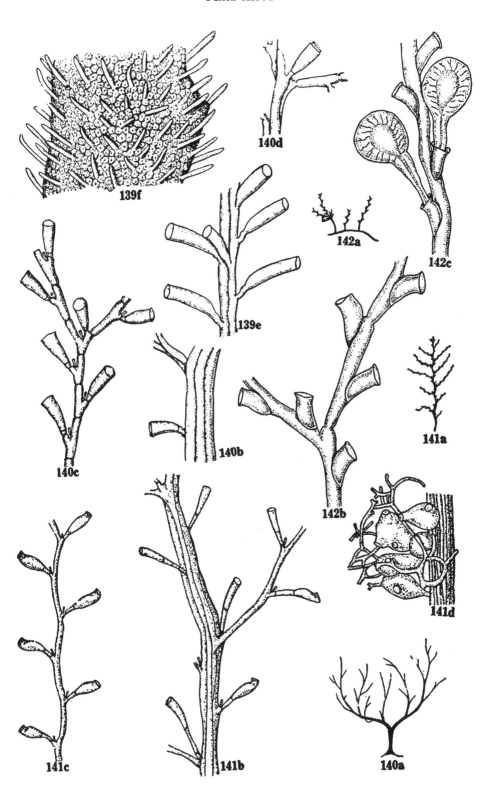

139f

140d

142c

140c

139e

142a

140b

142b

141a

141d

141c

141b

140a

PLATE XXVII

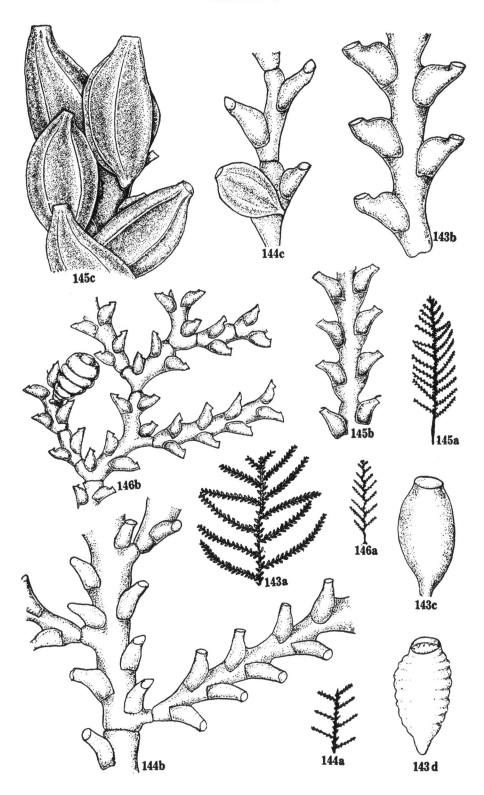

PLATE XXVIII

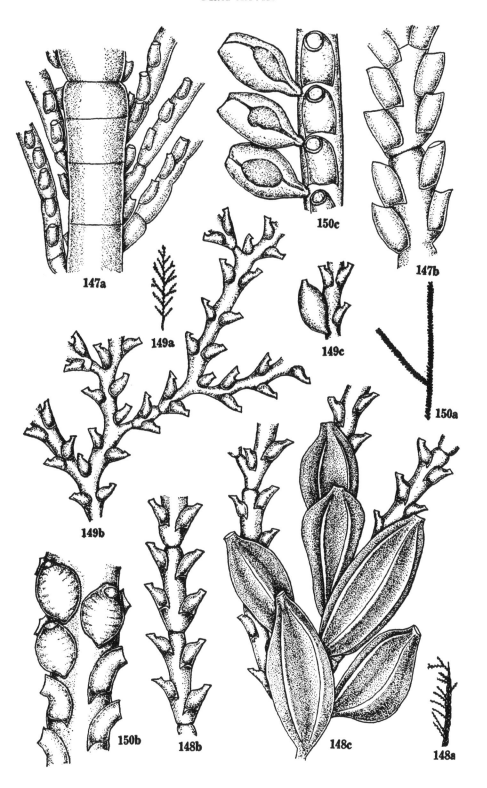

147a

150c

147b

149a

149c

150a

149b

150b

148b

148c

148a

PLATE XXIX

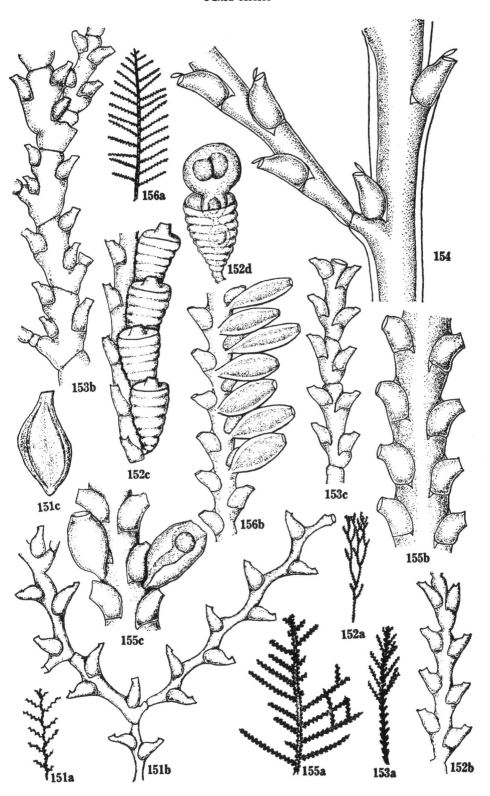

PLATE XXX

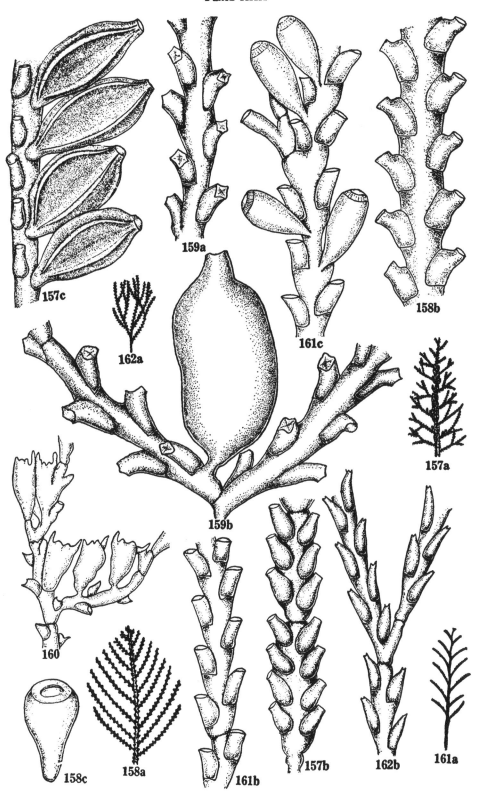

PLATE XXXI

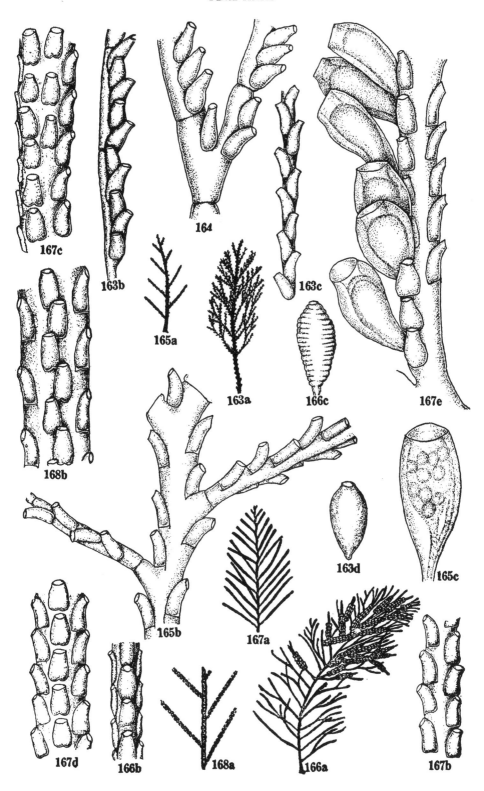

167c

163b

164

163c

167e

168b

165a

163a

166c

165b

163d

165c

167a

167d

166b

168a

166a

167b

PLATE XXXII

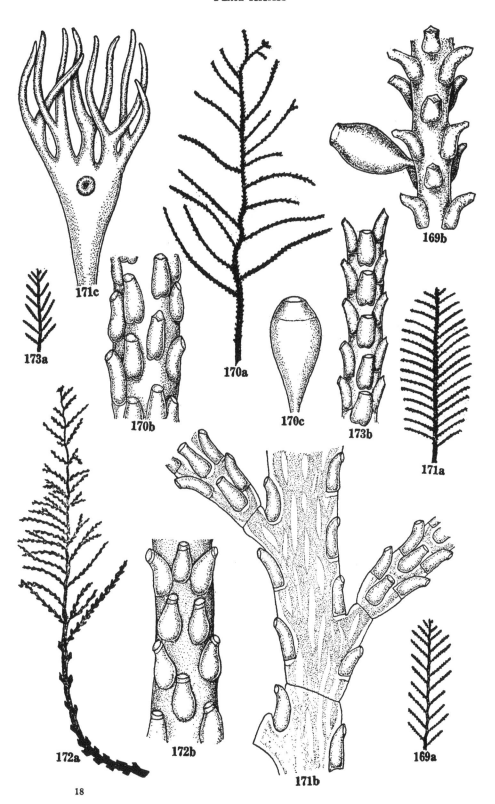

171c

173a

170b

170a

170c

173b

169b

171a

172a

172b

171b

169a

18

PLATE XXXIII

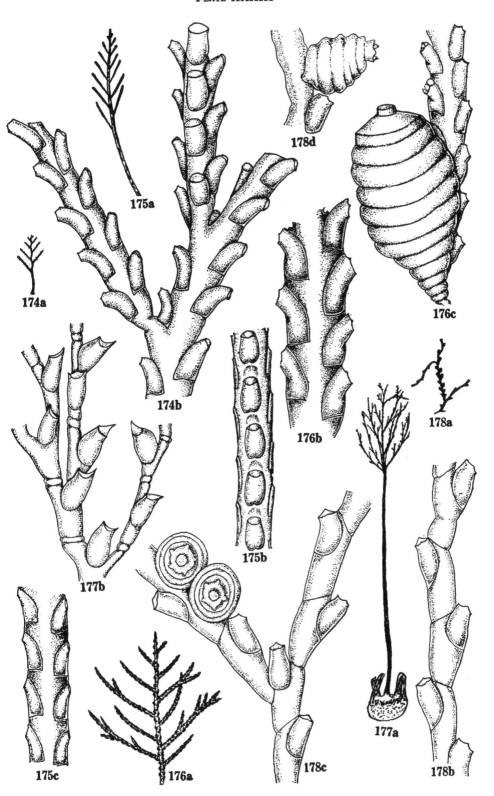

174a

175a

174b

175b

175c

176a

176b

176c

177a

177b

178a

178b

178c

178d

PLATE XXXIV

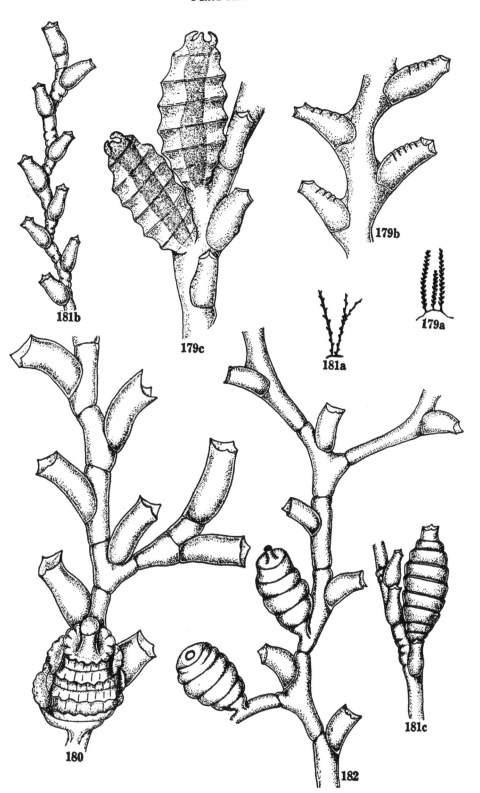

181b

179c

179b

181a

179a

180

182

181c

PLATE XXXV

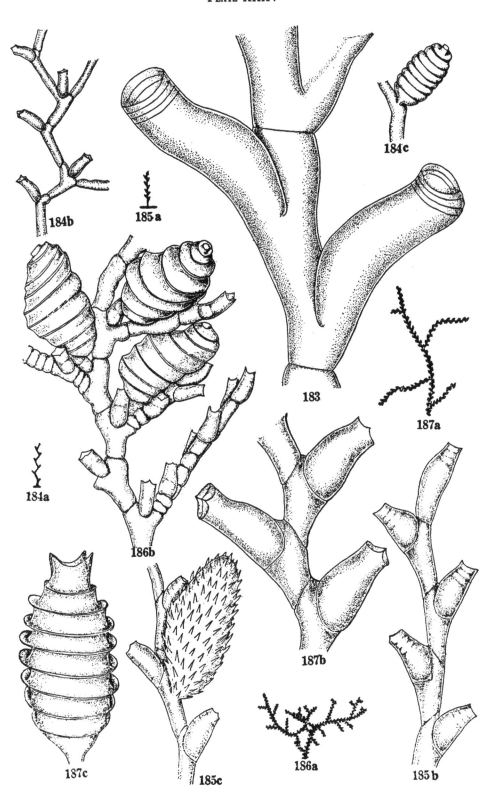

PLATE XXXVI

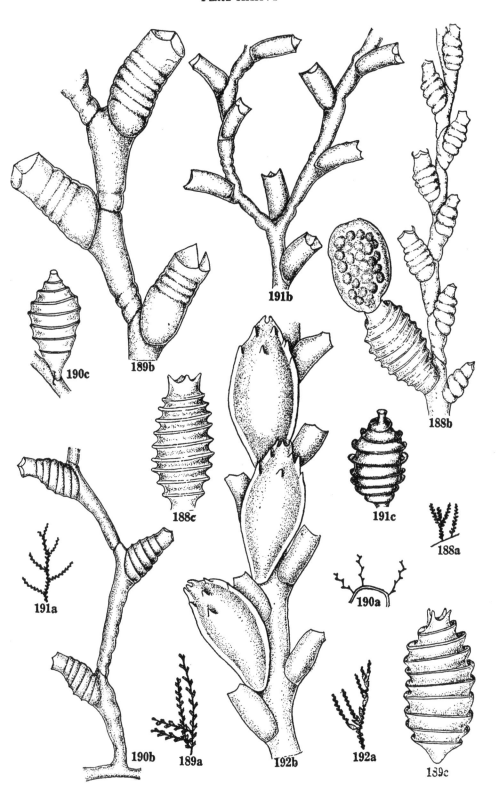

190c

189b

191b

188b

188c

191c

188a

191a

190a

190b   189a   192b   192a   189c

PLATE XXXVII

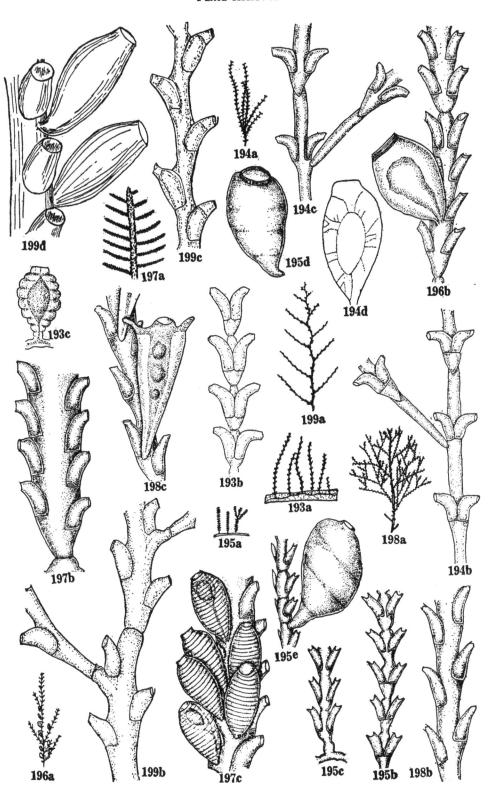

199d

197a

199c

194a

194c

195d

194d

196b

193c

198c

193b

199a

197b

195a

193a

198a

194b

196a

199b

197c

195e

195c

195b

198b

PLATE XXXVIII

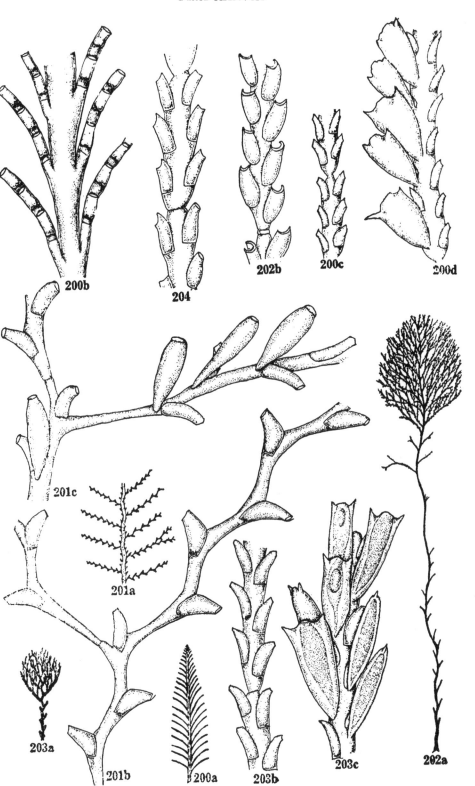

200b

204

202b

200c

200d

201c

201a

203a

201b

200a

203b

203c

202a

PLATE XXXIX

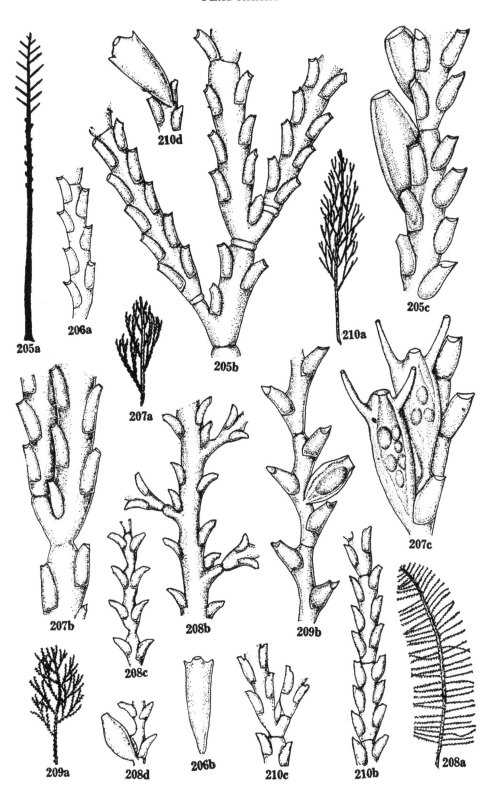

205a

206a

210d

205b

207a

210a

205c

207b

208c

208b

209b

207c

209a

208d

206b

210c

210b

208a

PLATE XL

Fig. 211. *Thuiaria thuja*
    a. Natural size.
    b. Portion of branch with hydrothecae.
    c. Gonangium.

Fig. 212. *Aglaophenia diegensis*
    a. Natural size.
    b. Portion of hydrocladium.
    c. Corbula.

Fig. 213. *Aglaophenia inconspicua*
    a. Natural size.
    b. Portion of hydrocladium.
    c. Corbula.

Fig. 214. *Aglaophenia latirostris*
    a. Natural size.
    b. Portion of hydrocladium.
    c. Corbula (After Nutting).

Fig. 215. *Aglaophenia lophocarpa*
    a. Natural size (After Allman).
    b. Portion of colony showing arrangement of hydrocladia and face view of hydrothecae (After Allman).
    c. Hydrothecae, lateral view (After Allman).
    d. Corbula (After Allman).

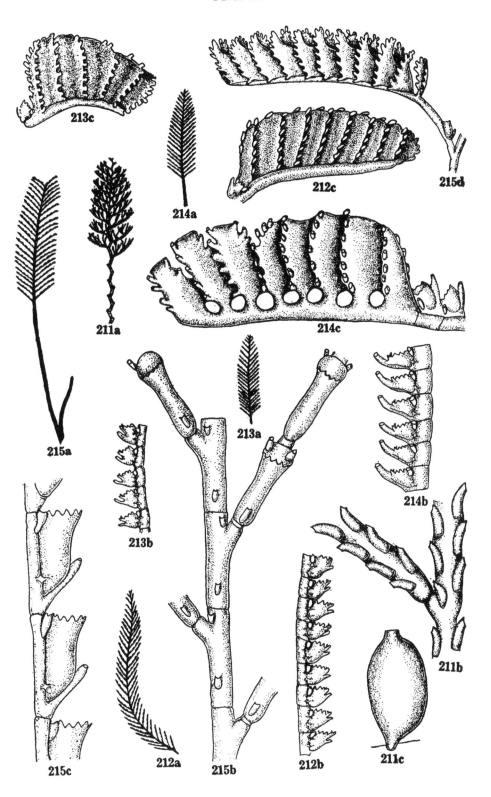

PLATE XL

213c

214a

215d

212c

211a

214c

215a

213a

214b

213b

215c

212a

215b

212b

211c

211b

PLATE XLI

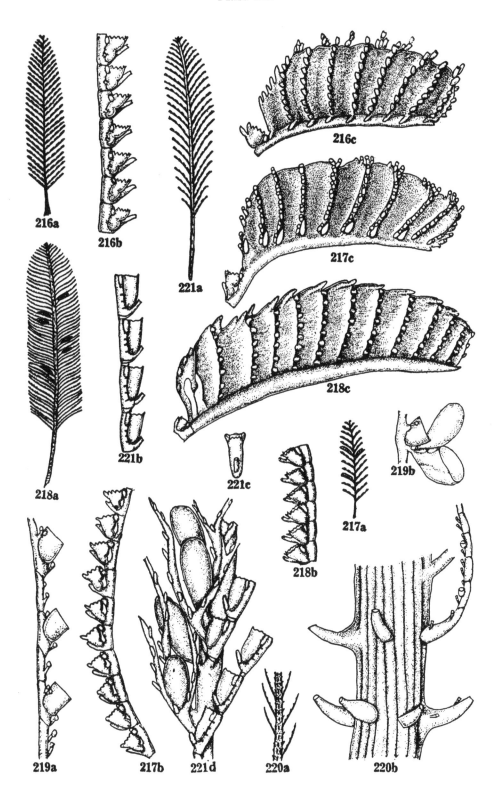

216a
216b
221a
216c
217c
218a
221b
218c
221c
217a
219b
218b
217b
221d
220a
220b
219a

PLATE XLII

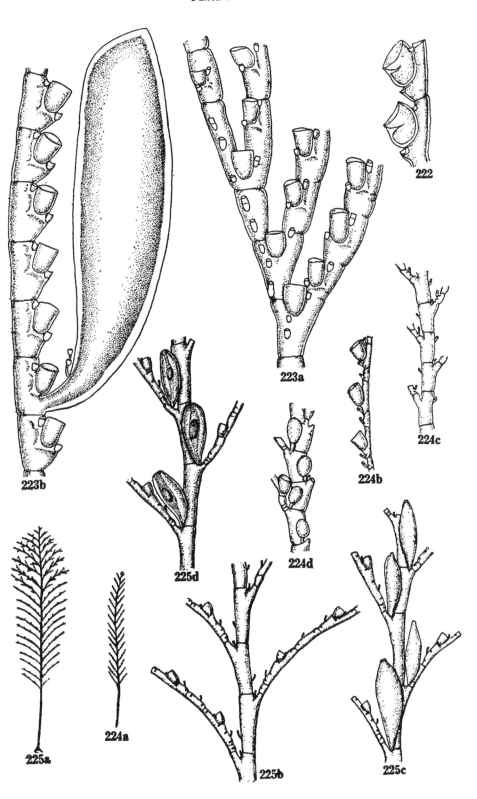

222

223a

223b

224a

224b

224c

224d

225a

225b

225c

225d

PLATE XLIII

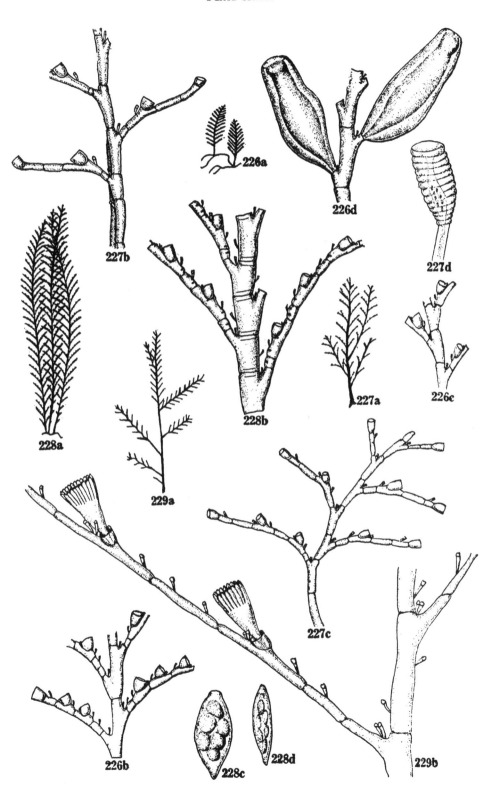

226a

226d

227b

227d

228a

228b

227a

226c

229a

227c

226b

228c 228d

229b

PLATE XLIV

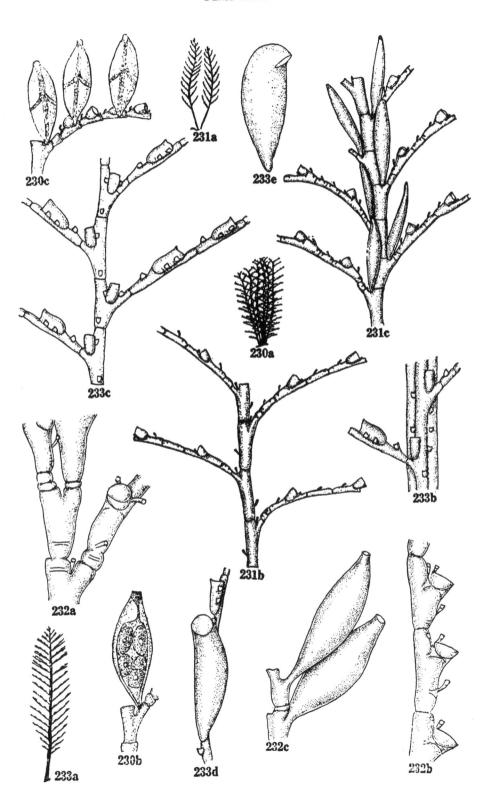